I0064903

# Wind Energy

Nicolette Fox

SYRAWOOD
PUBLISHING HOUSE

New York

Published by Syrawood Publishing House,
750 Third Avenue, 9th Floor,
New York, NY 10017, USA
www.syrawoodpublishinghouse.com

**Wind Energy**
Nicolette Fox

© 2022 Syrawood Publishing House

International Standard Book Number: 978-1-64740-125-2 (Hardback)

This book contains information obtained from authentic and highly regarded sources. All chapters are published with permission under the Creative Commons Attribution Share Alike License or equivalent. A wide variety of references are listed. Permissions and sources are indicated; for detailed attributions, please refer to the permissions page. Reasonable efforts have been made to publish reliable data and information, but the authors, editors and publisher cannot assume any responsibility for the validity of all materials or the consequences of their use.

**Trademark Notice:** Registered trademark of products or corporate names are used only for explanation and identification without intent to infringe.

**Cataloging-in-Publication Data**

Wind energy / Nicolette Fox.
    p. cm.
Includes bibliographical references and index.
ISBN 978-1-64740-125-2
1. Wind power. 2. Renewable energy sources. 3. Power resources. I. Fox, Nicolette.
TJ820 .W56 2022
333.92--dc23

# TABLE OF CONTENTS

# PREFACE

This book aims to help a broader range of students by exploring a wide variety of significant topics related to this discipline. It will help students in achieving a higher level of understanding of the subject and excel in their respective fields. This book would not have been possible without the unwavered support of my senior professors who took out the time to provide me feedback and help me with the process. I would also like to thank my family for their patience and support.

The kinetic energy of the wind is referred to as wind energy. It is used to provide wind power and involves the usage of air flow to generate electricity with the help of wind turbines. There are two types of modern wind turbines, namely horizontal axis and vertical axis wind turbines. The size of the wind turbines varies on the basis of their usage. Wind is a variable source of energy and can be used in conjunction with other forms of power generation such as hydropower and solar power to provide continuous power. There are various factors which determine the amount of energy which a turbine can harness from the wind. A few of these are wind speed, air density and swept area. This book is compiled in such a manner, that it will provide an in-depth knowledge about the concepts and applications of wind energy. It is appropriate for students seeking detailed information in this area as well as for experts. Coherent flow of topics, student-friendly language and extensive use of examples make this book an invaluable source of knowledge.

A brief overview of the book contents is provided below:

Chapter – What is Wind Energy?

The natural movement of air in the form of air current on the surface of Earth is known as wind. The kinetic energy of the wind is called as wind energy by which it generates electricity. The topics elaborated in this chapter will help in gaining a better perspective of wind and wind energy.

Chapter – Types of Wind Turbines

Wind turbines are the devices used for the conversion of kinetic energy of wind into electrical energy. The two main types of wind turbines are horizontal-axis turbines and vertical-axis turbines. This is an introductory chapter which will briefly introduce these types of wind turbines.

Chapter – Wind Energy Conversion Systems

Wind energy conversion system is used to convert the energy of wind movement into mechanical power which can be used to power machinery and electrical generator. It includes variable speed systems and grid connected systems. This chapter discusses these wind energy conversion systems in detail.

Chapter – Wind Power Calculations

The power output of the wind and wind turbine blade efficiency can be calculated by using Betz law, kinetic energy and potential energy equations. This chapter closely examines these laws and equations associated with wind power to provide an extensive understanding of the subject.

Chapter – Environmental Impacts of Wind Energy

Wind power can have adverse impacts on humans, wildlife and the environment. It includes degradation of the habitat of wildlife, plants and fish, threat to flying animals, noise pollution, visual impacts on landscape, etc. This chapter has been carefully written to provide an easy understanding of these impacts of wind power.

**Nicolette Fox**

# What is Wind Energy?

The natural movement of air in the form of air current on the surface of Earth is known as wind. The kinetic energy of the wind is called as wind energy by which it generates electricity. The topics elaborated in this chapter will help in gaining a better perspective of wind and wind energy.

## Wind

Wind is the movement of air relative to the surface of the Earth. Winds play a significant role in determining and controlling climate and weather.

Wind occurs because of horizontal and vertical differences (gradients) in atmospheric pressure. Accordingly, the distribution of winds is closely related to that of pressure. Near the Earth's surface, winds generally flow around regions of relatively low and high pressure—cyclones and anticyclones, respectively. They rotate counterclockwise around lows in the Northern Hemisphere and clockwise around those in the Southern Hemisphere. Similarly, wind systems rotate around the centres of highs in the opposite direction.

Wind bending palm trees Wind bending palm trees on a golf course.

In the middle and upper troposphere, the pressure systems are organized in a sequence of high-pressure ridges and low-pressure troughs, rather than in the closed, roughly circular systems nearer the surface of the Earth. They have a wavelike motion and interact to form a rather complex series of ridges and troughs. The largest of the wave patterns are the so-called standing waves that have three or four ridges and a corresponding number of troughs in a broad band in middle latitudes of the Northern Hemisphere. The westerlies of the Southern Hemisphere are much less strongly affected by standing disturbances. Associated with these long standing waves are the short waves (several hundred kilometres in wavelength) called traveling waves. Such traveling waves form the upper parts of near-surface cyclones and anticyclones to which they are linked, thus guiding their movement and development.

At high latitudes the winds are generally easterly near the ground. In low, tropical, and equatorial latitudes, the northeasterly trade winds north of the intertropical convergence zone (ICZ), or thermal equator, and the southeasterly trade winds south of the ICZ move toward the ICZ, which migrates north and south with the seasonal position of the Sun. Vertically, winds then rise and create towering cumulonimbus clouds and heavy rain on either side of the ICZ, which marks a narrow belt of near calms known as the doldrums. The winds then move poleward near the top of the troposphere before sinking again in the subtropical belts in each hemisphere. From here, winds again move toward the Equator as trade winds. These gigantic cells with overturning air in each of the hemispheres in low latitudes are known as the Hadley cells. In the mid-latitudes, oppositely rotating wind systems called Ferrel cells carry surface air poleward and upper tropospheric air toward the Hadley cells. The three-dimensional pattern of winds over the Earth, known as general circulation, is responsible for the fundamental latitudinal structure of pressure and air movement and, hence, of climates.

On a smaller scale are the local winds, systems that are associated with specific geographic locations and reflect the influence of topographic features. The most common of these local wind systems are the sea and land breezes, mountain and valley breezes, foehn winds (also called chinook, or Santa Ana, winds), and katabatic winds. Local winds exert a pronounced influence on local climate and are themselves affected by local weather conditions.

Wind speeds and gustiness are generally strongest by day when the heating of the ground by the Sun causes overturning of the air, the descending currents conserving the angular momentum of high-altitude winds. By night, the gustiness dies down and winds are generally lighter.

## Different Types of Wind

## Planetary Winds

The winds blowing throughout the year from one latitude to another in response to latitudinal differences in air pressure are called "planetary or prevailing winds". They involve large areas of the globe.

Two most important prevailing winds are trade winds and westerly winds.

## Trade Winds

These are extremely steady winds blowing from sub-tropical high pressure areas (30 °N and °S) towards the equatorial low pressure belt. These winds should have blown from the north to south in Northern Hemisphere and south to north in Southern Hemisphere, but, they get deflected to the right in Northern Hemisphere and to the left in Southern Hemisphere due to Coriolis effect and Ferrel's law. Thus, they blow as north eastern trades in Northern Hemisphere and south eastern trades in Southern Hemisphere.

They are also known as tropical easterlies, and they blow steadily in the same direction. They are noted for consistency in both force and direction.

## The Westerlies

These winds blow from sub tropical high pressure belts towards sub-polar low pressure belts. The westerlies of Southern Hemisphere are more stronger and constant in direction than Northern Hemisphere. These winds develop between 40° and 65 °S latitudes and these latitudes are known as Roaring Forties, Furious Fifties and Shrieking Sixties.

## Periodic Winds

Periodic winds change their direction periodically with the change in season, e.g., Monsoons, Land and Sea Breezes, Mountain and Valley Breezes.

## Monsoon Winds

These winds are seasonal winds and refer to wind systems that have a pronounced, seasonal reversal of direction. According to 'Flohn', monsoon is a seasonal modification of general Planetary Wind System.

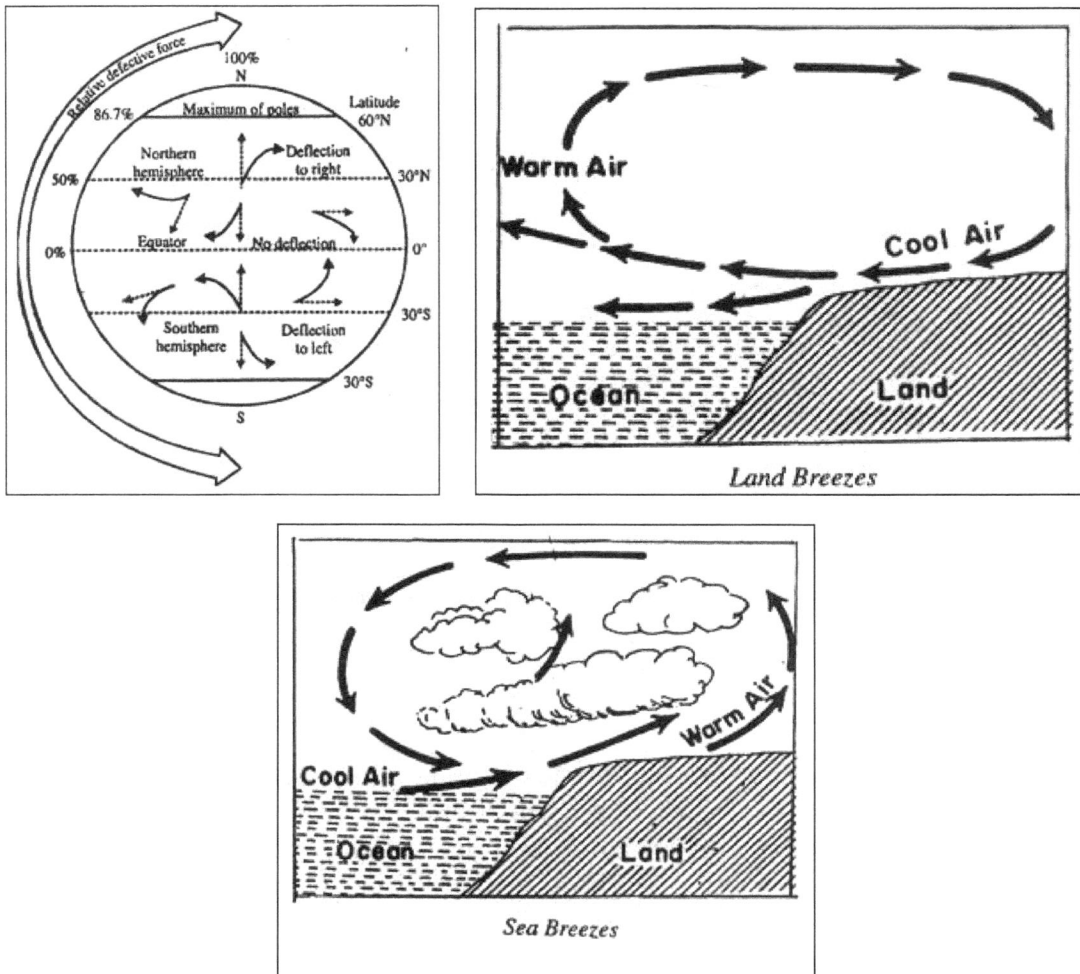

*Land Breezes*

*Sea Breezes*

Summer monsoon is called South Westerly Wind and is characterized by highly variable weather with frequent spells of drought and heavy rains. The winter monsoon is a gentle drift of air in which winds blow from the north-east and is known as North Easterly Wind.

## Land Breeze

At night, land masses cool quicker than sea due to rapid radiation which results in high pressure over land and low pressure over sea. And in calm, cloudless weather, air blows from land to sea. This breeze carries no moisture and is little warm and dry.

## Sea Breeze

In day time, the land being hotter than the sea develops low air pressure and the sea being cool develops high pressure. The air over land rises and is replaced by a cool breeze known as Sea Breeze from the sea, carrying some moisture.

## Mountain and Valley Breezes

A diurnal wind occurs in mountainous regions which are similar to Land and Sea Breezes. During the day the slopes of mountains are hot and air from valley flows up the slopes. This is known as "Valley Breeze". After sunset the pattern is reversed and cold air slides from mountain to valley and is called "mountain breeze".

## Local Winds

The local difference in temperature and pressure causes local winds. It is of four types: hot, cold, convectional and slope.

| Important Local Winds | |
|---|---|
| Local Wind | Area/Place of Blowing |
| Chinook. or Snoweather | U.S.A. and Canada |
| Foehn | The Alps and Switzerland |
| Samun | Iran |
| Norwester | New Zealand |
| Bergs | South Africa |
| Santa Anas | California (U.S.A.) |
| Sirocco | Sahara |
| Salano | Southern Spain |
| khamsin | Egypt |
| Simoon | Arabia |
| Gibli | Tunisia |
| Leveche | Sahara |
| Chili | Sahara |
| Brickfielder | Victoria (Australia) |
| Zonda | Argentina |
| Harmattan | Western Africa |
| Mistral | France |
| Bora | Former Yugoslavia |
| Tramontana | Northern Italy |
| Pampero | Argentina |
| Gregale | Sicily |
| Norther | Texas (U.S.A.) |
| Norta | Mexico |
| papagayo | Mexico |
| Loo | Northern India and Pakistan |

## Wind Energy

Wind energy is a converted form of solar energy which is produced by the nuclear fusion of hydrogen (H) into helium (He) in its core. The H → He fusion process creates heat and electromagnetic radiation streams out from the sun into space in all directions. Though only a small portion of solar radiation is intercepted by the earth, it provides almost all of earth's energy needs.

Wind energy represents a mainstream energy source of new power generation and an important player in the world's energy market. As a leading energy technology, wind power's technical maturity and speed of deployment is acknowledged, along with the fact that there is no practical upper limit to the percentage of wind that can be integrated into the electricity system. It has been estimated that the total solar power received by the earth is approximately $1.8 \times 10^{11}$ MW. Of this solar input, only 2% (i.e. $3.6 \times 10^9$ MW) is converted into wind energy and about 35% of wind energy is dissipated within 1000 m of the earth's surface. Therefore, the available wind power that can be converted into other forms of energy is approximately $1.26 \times 10^9$ MW. Because this value represents 20 times the rate of the present global energy consumption, wind energy in principle could meet entire energy needs of the world.

Compared with traditional energy sources, wind energy has a number of benefi ts and advantages. Unlike fossil fuels that emit harmful gases and nuclear power that generates radioactive wastes, wind power is a clean and environmentally friendly energy source. As an inexhaustible and free energy source, it is available and plentiful in most regions of the earth. In addition, more extensive use of wind power would help reduce the demands for fossil fuels, which may run out sometime in this century, according to their present consumptions. Furthermore, the cost per kWh of wind power is much lower than that of solar power.

Thus, as the most promising energy source, wind energy is believed to play a critical role in global power supply in the 21st century.

The rising concerns over global warming, environmental pollution, and energy security have increased interest in developing renewable and environmentally friendly energy sources such as wind, solar, hydropower, geothermal, hydrogen, and biomass as the replacements for fossil fuels. Wind energy can provide suitable solutions to the global climate change and energy crisis. The utilization of wind power essentially eliminates emissions of $CO_2$, $SO_2$, $NO_x$ and other harmful wastes as in traditional coal-fuel power plants or radioactive wastes in nuclear power plants. By further diversifying the energy supply, wind energy dramatically reduces the dependence on fossil fuels that are subject to price and supply instability, thus strengthening global energy security. During the recent three decades, tremendous growth in wind power has been seen all over the world. In 2009, the global annual installed wind generation capacity reached a record-breaking 37 GW, bringing the world total wind capacity to 158 GW. As the most promising renewable, clean, and reliable energy source, wind power is highly expected to take a much higher portion in power generation in the coming decades.

### Wind Generation

Wind results from the movement of air due to atmospheric pressure gradients. Wind flows from regions of higher pressure to regions of lower pressure. The larger the atmospheric pressure gradient, the higher the wind speed and thus, the greater the wind power that can be captured from the wind by means of wind energy-converting machinery.

The generation and movement of wind are complicated due to a number of factors. Among them, the most important factors are uneven solar heating, the Coriolis effect due to the earth's self-rotation, and local geographical conditions.

## Uneven Solar Heating

Among all factors affecting the wind generation, the uneven solar radiation on the earth's surface is the most important and critical one. The unevenness of the solar radiation can be attributed to four reasons.

First, the earth is a sphere revolving around the sun in the same plane as its equator. Because the surface of the earth is perpendicular to the path of the sunrays at the equator but parallel to the sunrays at the poles, the equator receives the greatest amount of energy per unit area, with energy dropping off toward the poles. Due to the spatial uneven heating on the earth, it forms a temperature gradient from the equator to the poles and a pressure gradient from the poles to the equator. Thus, hot air with lower air density at the equator rises up to the high atmosphere and moves towards the poles and cold air with higher density flows from the poles towards the equator along the earth's surface. Without considering the earth's self-rotation and the rotation-induced Coriolis force, the air circulation at each hemisphere forms a single cell, defined as the meridional circulation.

Second, the earth's self-rotating axis has a tilt of about 23.5° with respect to its ecliptic plane. It is the tilt of the earth's axis during the revolution around the sun that results in cyclic uneven heating, causing the yearly cycle of seasonal weather changes.

Third, the earth's surface is covered with different types of materials such as vegetation, rock, sand, water, ice/snow, etc. Each of these materials has different reflecting and absorbing rates to solar radiation, leading to high temperature on some areas (e.g. deserts) and low temperature on others (e.g. iced lakes), even at the same latitudes.

The fourth reason for uneven heating of solar radiation is due to the earth's topographic surface. There are a large number of mountains, valleys, hills, etc. on the earth, resulting in different solar radiation on the sunny and shady sides.

## Coriolis Force

The earth's self-rotation is another important factor to affect wind direction and speed. The Coriolis force, which is generated from the earth's self-rotation, deflects the direction of atmospheric movements. In the north atmosphere wind is deflected to the right and in the south atmosphere to the left. The Coriolis force depends on the earth's latitude; it is zero at the equator and reaches maximum values at the poles. In addition, the amount of deflection on wind also depends on the wind speed; slowly blowing wind is deflected only a small amount, while stronger wind deflected more.

In large-scale atmospheric movements, the combination of the pressure gradient due to the uneven solar radiation and the Coriolis force due to the earth's selfrotation causes the single meridional cell to break up into three convectional cells in each hemisphere: the Hadley cell, the Ferrel cell, and the Polar cell. Each cell has its own characteristic circulation pattern.

In the Northern Hemisphere, the Hadley cell circulation lies between the equator and north latitude 30°, dominating tropical and sub-tropical climates. The hot air rises at the equator and flows toward the North Pole in the upper atmosphere. This moving air is deflected by Coriolis force to create the northeast trade winds. At approximately north latitude 30°, Coriolis force becomes so strong to balance the pressure gradient force. As a result, the winds are defected to the west. The air accumulated at the upper atmosphere forms the subtropical high-pressure belt and thus sinks back to the earth's surface, splitting into two components: one returns to the equator to close the loop of the Hadley cell; another moves along the earth's surface toward North Pole to form the Ferrel Cell circulation, which lies between north latitude 30° and 60°. The air circulates toward the North Pole along the earth's surface until it collides with the cold air flowing from the North Pole at approximately north latitude 60°. Under the influence of Coriolis force, the moving air in this zone is deflected to produce westerlies. The Polar cell circulation lies between the North Pole and north latitude 60°. The cold air sinks down at the North Pole and flows along the earth's surface toward the equator. Near north latitude 60°, the Coriolis effect becomes significant to force the airflow to southwest.

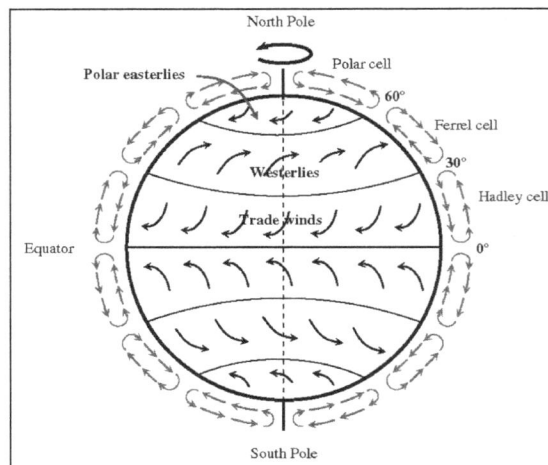

Idealized Atmospheric Circulations.

## Local Geography

The roughness on the earth's surface is a result of both natural geography and manmade structures. Frictional drag and obstructions near the earth's surface generally retard with wind speed and induce a phenomenon known as wind shear. The rate at which wind speed increases with height varies on the basis of local conditions of the topography, terrain, and climate, with the greatest rates of increases observed over the roughest terrain. A reliable approximation is that wind speed increases about 10% with each doubling of height.

In addition, some special geographic structures can strongly enhance the wind intensity. For instance, wind that blows through mountain passes can form mountain jets with high speeds.

## Wind Energy Applications

The use of wind energy can be traced back thousands of years to many ancient civilizations. The ancient human histories have revealed that wind energy was discovered and used independently at several sites of the earth.

## Sailing

As early as about 4000 B.C., the ancient Chinese were the first to attach sails to their primitive rafts. In Han Dynasty, Chinese junks were developed and used as ocean-going vessels. As recorded in a book wrote in the third century, there were multi-mast, multi-sail junks sailing in the South Sea, capable of carrying 700 people with 260 tons of cargo. Two ancient Chinese junks are shown in figure. Figure (a) is a two-mast Chinese junk ship for shipping grain, quoted from the famous encyclopedic science and technology book Exploitation of the works of nature. Figure (b) illustrates a wheel boat in Song Dynasty. It mentioned in that this type of wheel boats was used during the war between Song and Jin Dynasty.

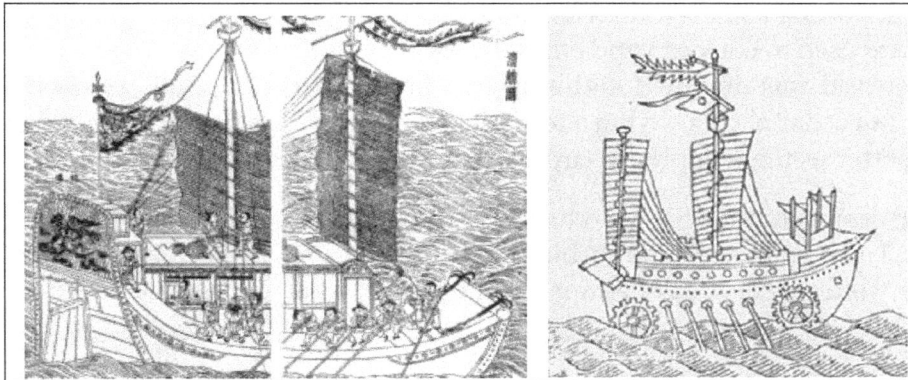

Ancient Chinese junks (ships): (a) two-mast junk ship; (b) wheel boat.

Approximately at 3400 BC, the ancient Egyptians launched their first sailing vessels initially to sail on the Nile River, and later along the coasts of the Mediterranean. Around 1250 BC, Egyptians built fairly sophisticated ships to sail on the Red Sea. The wind-powered ships had dominated water transport for a long time until the invention of steam engines in the 19th century.

## Wind in Metal Smelting Processes

About 300 BC, ancient Sinhalese had taken advantage of the strong monsoon winds to provide furnaces with sufficient air for raising the temperatures inside furnaces in excess of 1100 °C in iron smelting processes. This technique was capable of producing high-carbon steel.

The double acting piston bellows was invented in China and was widely used in metallurgy in the fourth century BC. It was the capacity of this type of bellows to deliver continuous blasts of air into furnaces to raise high enough temperatures for smelting iron. In such a way, ancient Chinese could once cast several tons of iron.

## Windmills

China has long history of using windmills. The unearthed mural paintings from the tombs of the late Eastern Han Dynasty at Sandaohao, Liaoyang City, have shown the exquisite images of windmills, evidencing the use of windmills in China for at least approximately 1800 years.

The practical vertical axis windmills were built in Sistan (eastern Persia) for grain grinding and water pumping, as recorded by a Persian geographer in the ninth century.

The horizontal axis windmills were invented in northwestern Europe in 1180s. The earlier windmills typically featured four blades and mounted on central posts – known as Post mill. Later, several types of windmills, e.g. Smock mill, Dutch mill, and Fan mill, had been developed in the Netherlands and Denmark, based on the improvements on Post mill.

The horizontal axis windmills have become dominant in Europe and North America for many centuries due to their higher operation efficiency and technical advantages over vertical axis windmills.

## Wind Turbines

Unlike windmills which are used directly to do work such as water pumping or grain grinding, wind turbines are used to convert wind energy to electricity. The first automatically operated wind turbine in the world was designed and built by Charles Brush in 1888. This wind turbine was equipped with 144 cedar blades having a rotating diameter of 17 m. It generated a peak power of 12 kW to charge batteries that supply DC current to lamps and electric motors.

As a pioneering design for modern wind turbines, the Gedser wind turbine was built in Denmark in the mid 1950s. Today, modern wind turbines in wind farms have typically three blades, operating at relative high wind speeds for the power output up to several megawatts.

## Kites

Kites were invented in China as early as the fifth or fourth centuries BC. A famous Chinese ancient legalist Han Fei-Zi mentioned in his book that an ancient philosopher Mo Ze spent three years to make a kite with wood but failed after one-day flight.

## Wind Energy Characteristics

Wind energy is a special form of kinetic energy in air as it flows. Wind energy can be either converted into electrical energy by power converting machines or directly used for pumping water, sailing ships, or grinding gain.

## Wind Power

Kinetic energy exists whenever an object of a given mass is in motion with a translational or rotational speed. When air is in motion, the kinetic energy in moving air can be determined as:

$$E_k = \frac{1}{2} m \bar{u}^2$$

where m is the air mass and $\bar{u}$ is the mean wind speed over a suitable time period. The wind power can be obtained by differentiating the kinetic energy in wind with respect to time, i.e.

$$P_W = \frac{dE_k}{dt} = \frac{1}{2} \dot{m} \bar{u}^2$$

However, only a small portion of wind power can be converted into electrical power. When wind passes through a wind turbine and drives blades to rotate, the corresponding wind mass flowrate is:

$$\dot{m} = \rho A \bar{u}^3$$

where $\rho$ is the air density and A is the swept area of blades. Substituting the equation, the available power in wind $P_w$ can be expressed as:

$$P_W = \frac{1}{2} \rho A \bar{u}^3$$

An examination of eqn $P_W = \frac{1}{2} \rho A \bar{u}^3$ reveals that in order to obtain a higher wind power, it requires a higher wind speed, a longer length of blades for gaining a larger swept area, and a higher air density. Because the wind power output is proportional to the cubic power of the mean wind speed, a small variation in wind speed can result in a large change in wind power.

## Blade Swept Area

As shown in figure, the blade swept area can be calculated from the formula:

$$A = \pi \left[ (1+r)^2 - r^2 \right] = \pi l (l + 2r)$$

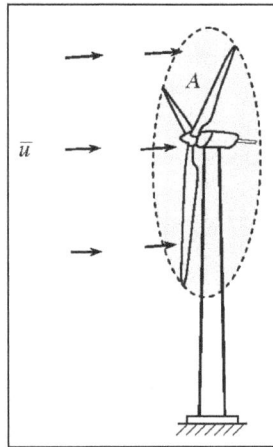

Swept area of wind turbine blades.

where $l$ is the length of wind blades and r is the radius of the hub. Thus, by doubling the length of wind blades, the swept area can be increased by the factor up to 4. When $l \gg 2r, A \approx \pi l^2$.

## Air Density

Another important parameter that directly affects the wind power generation is the density of air, which can be calculated from the equation of state:

$$\rho = \frac{p}{RT}$$

where $\rho$ is the local air pressure, R is the gas constant (287 J/kg-K for air), and T is the local air temperature in K.

The hydrostatic equation states that whenever there is no vertical motion, the difference in pressure between two heights is caused by the mass of the air layer:

$$\mathrm{d}p = -\rho g\, \mathrm{d}z$$

where g is the acceleration of gravity. Combining eqns $\rho = \dfrac{p}{RT}$ and $\mathrm{d}p = -\rho g\, \mathrm{d}z$, yields:

$$\frac{\mathrm{d}p}{p} = -\frac{g}{RT}\mathrm{d}z$$

The acceleration of gravity g decreases with the height above the earth's surface $z$:

$$g = g_0\left(1 - \frac{4z}{D}\right)$$

where g0 is the acceleration of gravity at the ground and D is the diameter of the earth. However, for the acceleration of gravity g, the variation in height can be ignored because D is much larger than $4z$.

In addition, temperature is inversely proportional to the height. Assume that $dT/dz = c$, it can be derived that:

$$p = p_0\left(\frac{T}{T_0}\right)^{-g/cR}$$

where p0 and T0 are the air pressure and temperature at the ground, respectively. Combining eqns

$$\rho = \frac{p}{RT} \text{ and } p = p_0\left(\frac{T}{T_0}\right)^{-g/cR}, \text{ it gives}$$

This equation indicates that the density of air decreases nonlinearly with the height above the sea level.

## Wind Power Density

Wind power density is a comprehensive index in evaluating the wind resource at a particular site. It is the available wind power in airflow through a perpendicular cross-sectional unit area in a unit time period. The classes of wind power density at two standard wind measurement heights are listed in table.

Some of wind resource assessments utilize 50 m towers with sensors installed at intermediate levels (10 m, 20 m, etc.). For large-scale wind plants, class rating of 4 or higher is preferred.

Table: Classes of wind power density.

| | 10 m height | | 50 m height | |
|---|---|---|---|---|
| Wind power Class | Wind power density (W/m²) | Mean wind speed (m/s)) | Wind powe rdensity(W/m²) | Mean wind speed (m/s) |
| 1 | <100 | <4.4 | <200 | <5.6 |

| 2 | 100–150 | 4.4–5.1 | 200–300 | 5.6–6.4 |
|---|---|---|---|---|
| 3 | 100–150 | 5.1–5.6 | 300–400 | 6.4–7.0 |
| 4 | 200–250 | 5.6–6.0 | 400–500 | 7.0–7.5 |
| 5 | 250–300 | 6.0–6.4 | 500–600 | 7.5–8.0 |
| 6 | 300–350 | 6.4–7.0 | 600–800 | 8.0–8.8 |
| 7 | >400 | >7.0 | >800 | >8.8 |

## Wind Characteristics

Wind varies with the geographical locations, time of day, season, and height above the earth's surface, weather, and local landforms. The understanding of the wind characteristics will help optimize wind turbine design, develop wind measuring techniques, and select wind farm sites.

## Wind Speed

Wind speed is one of the most critical characteristics in wind power generation. In fact, wind speed varies in both time and space, determined by many factors such as geographic and weather conditions. Because wind speed is a random parameter, measured wind speed data are usually dealt with using statistical methods.

The diurnal variations of average wind speeds are often described by sine waves. As an example, the diurnal variations of hourly wind speed values, which are the average values calculated based on the data between 1970 and 1984, at Dhahran, Saudi Arabia have shown the wavy pattern. The wind speeds are higher in daytime and the maximum speed occurs at about 3 p.m., indicating that the daytime wind speed is proportional to the strength of sunlight. George reported that wind speed at Lubbock, TX is near constant during dark hours, and follows a curvilinear pattern during daylight hours. Later, George have demonstrated that diurnal wind patterns at five locations in the Great Plains follow a pattern similar to that observed in.

Based on the wind speed data for the period 1970–2003 from up to 66 onshore sites around UK, Sinden has concluded that monthly average wind speed is inversely propositional to the monthly average temperature, i.e. it is higher in the winter and lower in the summer. The maximum wind speed occurs in January and the minimum in August. Hassanm and Hill have reported that the month-to-month variation of mean wind speed values over the period of 1970–1984 at Dhahran, Saudi Arabia has shown the wavy pattern. However, because the variation in temperature at Dhahran is small over the whole year, there is no a clear correlation between wind speed and temperatures.

The year-to-year variation of yearly mean wind speeds depends highly on selected locations and thus there is no common correlation to predict it. For instance, except for several years, the annual mean wind speeds decrease all the way from 1970 to 1983 at Dhahran, Saudi Arabia. In UK, this variation displays in a more fluctuated matter for the period 1970–2003. Similarly, a significant variation in the annual mean wind speed over 20-year period is reported in, with maximum and minimum values ranging from less than 7.8 to nearly 9.2 m/s. The long-term wind data obtained from automated synoptic observation system of meteorological observatories were analyzed and reported by Ko. The results show that fluctuation in yearly average wind speed occurs at the observed sites; it tends to slightly decrease at Jeju Island, while the other two sites have random trends.

## Weibull Distribution

The variation in wind speed at a particular site can be best described using the Weibull distribution function, which illustrates the probability of different mean wind speeds occurring at the site during a period of time. The probability density function of a Weibull random variable $\bar{u}$ is:

$$f\left(\bar{u},k,\lambda\right)=\begin{cases}\dfrac{k}{\lambda}\left(\dfrac{\bar{u}}{\lambda}\right)^{k-1}\exp\left(-\left(\dfrac{\bar{u}}{\lambda}\right)^{k}\right) & \bar{u}\geq0\\ 0 & \bar{u}\leq0\end{cases}$$

where l is the scale factor which is closely related to the mean wind speed and k is the shape factor which is a measurement of the width of the distribution. These two parameters can be determined from the statistical analysis of measured wind speed data at the site. It has been reported that Weibull distribution can give good fits to observed wind speed data. As an example, the Weibull distributions for various mean wind speeds are displayed.

## Wind Turbulence

Wind turbulence is the fluctuation in wind speed in short time scales, especially for the horizontal velocity component. The wind speed u( t ) at any instant time t can be considered as having two components: the mean wind speed $\bar{u}$ and the instantaneous speed fluctuation u'( t ), i.e.:

$$u\left(t\right)=\bar{u}+u'\left(t\right)$$

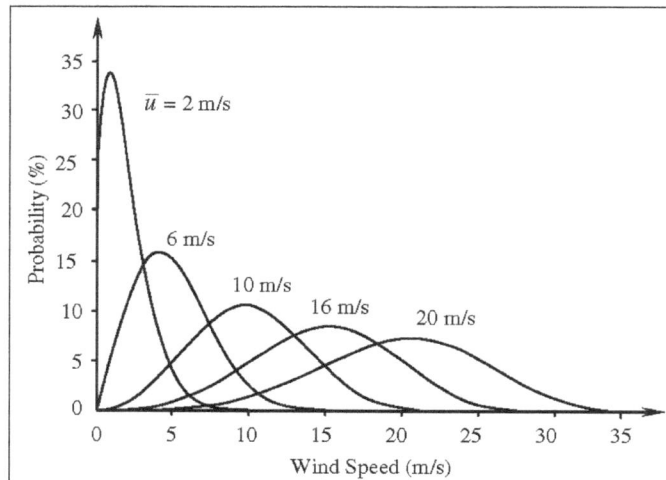

Weibull distributions for various mean wind speeds.

Wind turbulence has a strong impact on the power output fluctuation of wind turbine. Heavy turbulence may generate large dynamic fatigue loads acting on the turbine and thus reduce the expected turbine lifetime or result in turbine failure.

In selection of wind farm sites, the knowledge of wind turbulence intensity is crucial for the stability of wind power production. The wind turbulence intensity I is defined as the ratio of the standard deviation $\sigma_u$ to the mean wind velocity $\bar{u}$ :

$$I = \frac{\sigma_u}{\bar{u}}$$

where both $\sigma_u$ and $\bar{u}$ are measured at the same point and averaged over the same period of time.

## Wind Gust

Wind gust refers to a phenomenon that a wind blasts with a sudden increase in wind speed in a relatively small interval of time. In case of sudden turbulent gusts, wind speed, turbulence, and wind shear may change drastically. Reducing rotor imbalance while maintaining the power output of wind turbine generator constant during such sudden turbulent gusts calls for relatively rapid changes of the pitch angle of the blades. However, there is typically a time lag between the occurrence of a turbulent gust and the actual pitching of the blades based upon dynamics of the pitch control actuator and the large inertia of the mechanical components. As a result, load imbalances and generator speed, and hence oscillations in the turbine components may increase considerably during such turbulent gusts, and may exceed the maximum prescribed power output level. Moreover, sudden turbulent gusts may also significantly increase tower fore-aft and side-to-side bending moments due to increase in the effect of wind shear.

To ensure safe operation of wind farms, wind gust predictions are highly desired. Several different gust prediction methods have been proposed. Contrary to most techniques used in operational weather forecasting, Brasseur developed a new wind gust prediction method based on physical consideration. In another study, it reported that using a gust factor, which is defined as peak gust over the mean wind speed, could well forecast wind gust speeds. These results are in agreement with previous work by other investigators.

## Wind Direction

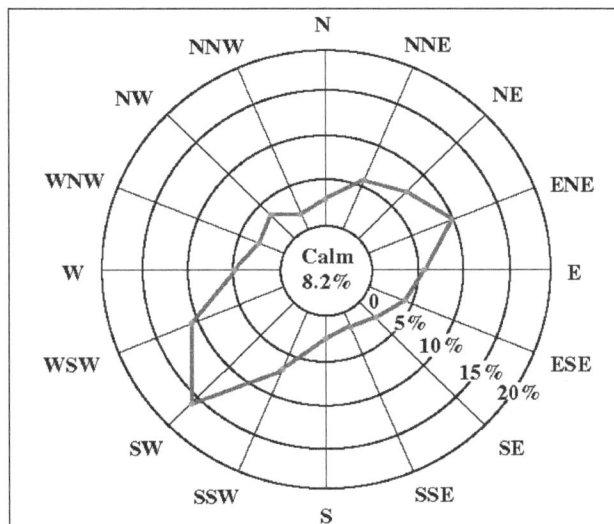

Wind rose diagram for wind directions.

Wind direction is one of the wind characteristics. Statistical data of wind directions over a long period of time is very important in the site selection of wind farm and the layout of wind turbines in the wind farm.

The wind rose diagram is a useful tool of analyzing wind data that are related to wind directions at a particular location over a specific time period (year, season, month, week, etc.). This circular diagram displays the relative frequency of wind directions in 8 or 16 principal directions. As an example shown in figure, there are 16 radial lines in the wind rose diagram, with 22.5° apart from each other. The length of each line is proportional to the frequency of wind direction. The frequency of calm or near calm air is given as a number in the central circle. Some wind rose diagrams may also contain the information of wind speeds.

## Wind Shear

Wind shear is a meteorological phenomenon in which wind increases with the height above the ground. The effect of height on the wind speed is mainly due to roughness on the earth's surface and can be estimated using the Hellmann power equation that relates wind speeds at two different heights:

$$u(z) = u(z_0)\left(\frac{z}{z_0}\right)^a$$

where $z$ is the height above the earth's surface, $z_0$ is the reference height for which wind speed $u(z_0)$ is known, and a is the wind shear coefficient. In practice, a depends on a number of factors, including the roughness of the surrounding landscape, height, time of day, season, and locations. The wind shear coefficient is generally lower in daytime and higher at night. Empirical results indicate that wind shear often follows the "1/7 power law" (i.e. a = 1/7). The values of wind shear coefficient for different surface roughness are provided in.

Because the power output of wind turbine strongly depends on the wind speed at the hub height, modern wind turbines are built at the height greater than 80 m, for capturing more wind energy and lowering cost per unit power output.

## Modern Wind Turbines

A modern wind turbine is an energy-converting machine to convert the kinetic energy of wind into mechanical energy and in turn into electrical energy. In the recent three decades, remarkable advances in wind turbine design have been achieved along with modern technological developments. It has been estimated that advances in aerodynamics, structural dynamics, and micrometeorology may contribute to a 5% annual increase in the energy yield of wind turbines.

Various wind turbine concepts have been developed and built for maximizing the wind energy output, minimizing the turbine cost, and increasing the turbine efficiency and reliability.

## Wind Turbine Classification

Wind turbines can be classified according to the turbine generator configuration, airflow path relatively to the turbine rotor, turbine capacity, the generator-driving pattern, the power supply mode, and the location of turbine installation.

## Horizontal-axis and Vertical-axis Wind Turbines

When considering the configuration of the rotating axis of rotor blades, modern wind turbines can be classified into the horizontal-axis and vertical-axis turbines.

Most commercial wind turbines today belong to the horizontal-axis type, in which the rotating axis of blades is parallel to the wind stream. The advantages of this type of wind turbines include the high turbine efficiency, high power density, low cut-in wind speeds, and low cost per unit power output.

Several typical vertical-axis wind turbines are shown in figure. The blades of the vertical-axis wind turbines rotate with respect to their vertical axes that are perpendicular to the ground. A significant advantage of vertical-axis wind turbine is that the turbine can accept wind from any direction and thus no yaw control is needed. Since the wind generator, gearbox, and other main turbine components can be set up on the ground, it greatly simplifies the wind tower design and construction, and consequently reduces the turbine cost. However, the vertical-axis wind turbines must use an external energy source to rotate the blades during initialization. Because the axis of the wind turbine is supported only on one end at the ground, its maximum practical height is thus limited. Due to the lower wind power efficiency, vertical-axis wind turbines today make up only a small percentage of wind turbines.

## Upwind and Downwind Wind Turbines

Based on the configuration of the wind rotor with respect to the wind flowing direction, the horizontal-axis wind turbines can be further classified as upwind and downwind wind turbines. The majority of horizontal-axis wind turbines being used today are upwind turbines, in which the wind rotors face the wind. The main advantage of upwind designs is to avoid the distortion of the flow field as the wind passes though the wind tower and nacelle.

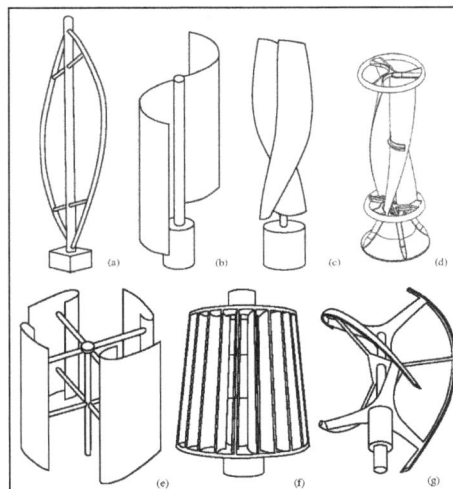

Several typical types of vertical-axis wind turbines: (a) Darrius; (b) Savonius; (c) Solarwind; (d) Helical; (e) Noguchi; (f) Maglev; (g) Cochrane.

For a downwind turbine, wind blows first through the nacelle and tower and then the rotor blades. This configuration enables the rotor blades to be made more flexible without considering tower strike. However, because of the influence of the distorted unstable wakes behind the tower and nacelle, the wind power output generated from a downwind turbine fluctuates greatly. In addition, the unstable flow field may result in more aerodynamic losses and introduce more fatigue loads on the turbine. Furthermore, the blades in a downwind wind turbine may produce higher impulsive or thumping noise.

## Wind Turbine Capacity

Wind turbines can be divided into a number of broad categories in view of their rated capacities: micro, small, medium, large, and ultra-large wind turbines. Though a restricted definition of micro wind turbines is not available, it is accepted that a turbine with the rated power less than several kilowatts can be categorized as micro wind turbine. Micro wind turbines are especially suitable in locations where the electrical grid is unavailable. They can be used on a per-structure basis, such as street lighting, water pumping, and residents at remote areas, particularly in developing countries. Because micro wind turbines need relatively low cut-in speeds at start-up and operate in moderate wind speeds, they can be extensively installed in most areas around the world for fully utilizing wind resources and greatly enhancing wind power generation availability.

Small wind turbines usually refer to the turbines with the output power less than 100 kW. Small wind turbines have been extensively used at residential houses, farms, and other individual remote applications such as water pumping stations, telecom sites, etc., in rural regions. Distributed small wind turbines can increase electricity supply in the regions while delaying or avoiding the need to increase the capacity of transmission lines.

The most common wind turbines have medium sizes with power ratings from 100 kW to 1 MW. This type of wind turbines can be used either on-grid or off-grid systems for village power, hybrid systems, distributed power, wind power plants, etc.

Megawatt wind turbines up to 10 MW may be classified as large wind turbines. In recent years, multi-megawatt wind turbines have become the mainstream of the international wind power market. Most wind farms presently use megawatt wind turbines, especially in offshore wind farms.

Ultra-large wind turbines are referred to wind turbines with the capacity more than 10 MW. This type of wind turbine is still in the earlier stages of research and development.

## Direct Drive and Geared Drive Wind Turbines

According to the drivetrain condition in a wind generator system, wind turbines can be classifi ed as either direct drive or geared drive groups. To increase the generator rotor rotating speed to gain a higher power output, a regular geared drive wind turbine typically uses a multi-stage gearbox to take the rotational speed from the low-speed shaft of the blade rotor and transform it into a fast rotation on the high-speed shaft of the generator rotor. The advantages of geared generator systems include lower cost and smaller size and weight. However, utilization of a gearbox can significantly lower wind turbine reliability and increase turbine noise level and mechanical losses.

By eliminating the multi-stage gearbox from a generator system, the generator shaft is directly connected to the blade rotor. Therefore, the direct-drive concept is more superior in terms of energy efficiency, reliability, and design simplicity.

## On-grid and Off-grid Wind Turbines

Wind turbines can be used for either on-grid or off-grid applications. Most medium-size and almost all large-size wind turbines are used in grid tied applications. One of the obvious advantages for on-grid wind turbine systems is that there is no energy storage problem.

As the contrast, most of small wind turbines are off-grid for residential homes, farms, telecommunications, and other applications. However, as an intermittent power source, wind power produced from off-grid wind turbines may change dramatically over a short period of time with little warning. Consequently, off-grid wind turbines are usually used in connection with batteries, diesel generators, and photovoltaic systems for improving the stability of wind power supply.

## Onshore and Offshore Wind Turbines

Onshore wind turbines have a long history on its development. There are a number of advantages of onshore turbines, including lower cost of foundations, easier integration with the electrical-grid network, lower cost in tower building and turbine installation, and more convenient access for operation and maintenance.

Offshore wind turbines have developed faster than onshore since the 1990s due to the excellent offshore wind resource, in terms of wind power intensity and continuity. A wind turbine installed offshore can make higher power output and operate more hours each year compared with the same turbine installed onshore. In addition, environmental restrictions are more lax at offshore sites than at onshore sites. For instance, turbine noise is no long an issue for offshore wind turbines.

## Wind Turbine Configuration

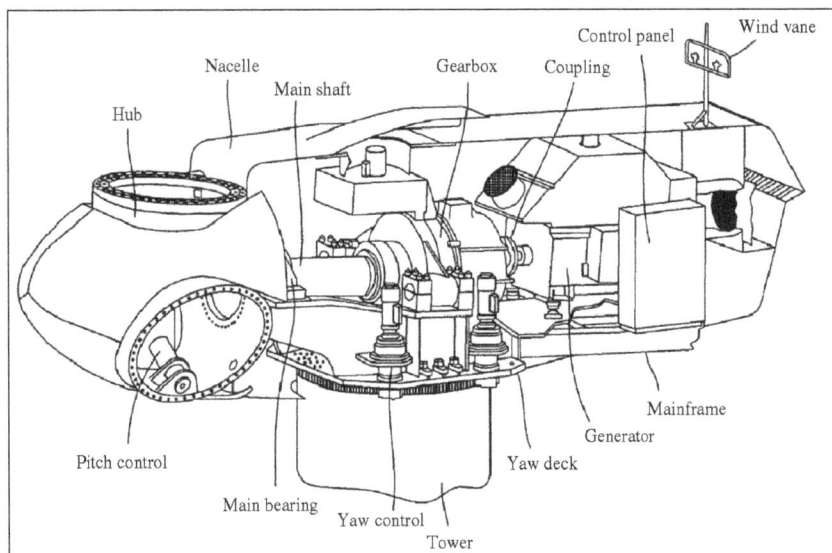

A horizontal-axis wind turbine configuration.

Most of the modern large wind turbines are horizontal-axis turbines with typically three blades. As shown in Fig, a wind turbine is comprised of a nacelle, which is positioned on the top of a wind tower, housing the most turbine components inside. Three blades (not shown) mounted on the rotor hub, which is connected via the main shaft to the gearbox. The rotor of the wind generator is connected to the output shaft of the gearbox. Thus, the slow rotating speed of the rotor hub is increased to a desired high rotating speed of the generator rotor.

Using the pitch control system, each blade is pitched individually to optimize the angle of attack of the blade for allowing a higher energy capture in normal operation and for protecting the turbine components (blade, tower, etc.) from damaging in emergency situations. With the feedback information such as measured instantaneous wind direction and speed from the wind vane, the yaw control system provides the yaw orientation control for ensuring the turbine constantly against the wind.

## Wind Power Parameters

## Power Coefficient

The conversion of wind energy to electrical energy involves primarily two stages: in the first stage, kinetic energy in wind is converted into mechanical energy to drive the shaft of a wind generator. The critical converting devices in this stage are wind blades. For maximizing the capture of wind energy, wind blades need to be carefully designed.

The power coefficient $C_p$ deals with the converting efficiency in the first stage, defined as the ratio of the actually captured mechanical power by blades to the available power in wind:

$$C_p = \frac{P_{me,out}}{P_w} = \frac{P_{me,out}}{(1/2)\rho A \bar{u}^3}$$

Because there are various aerodynamic losses in wind turbine systems, for instance, blade-tip, blade-root, profile, and wake rotation losses, etc., the real power coefficient $C_p$ is much lower than its theoretical limit, usually ranging from 30 to 45%.

## Total Power Conversion Coefficient and Effective Power Output

In the second stage, mechanical energy captured by wind blades is further converted into electrical energy via wind generators. In this stage, the converting efficiency is determined by several parameters.

- Gearbox efficiency $\eta_{gear}$ – The power losses in a gearbox can be classified as load-dependent and no-load power losses. The load-dependent losses consist of gear tooth friction and bearing losses and no-load losses consist of oil churning, windage, and shaft seal losses. The planetary gearboxes, which are widely used in wind turbines, have higher power transmission efficiencies over traditional gearboxes.

- Generator efficiency $\eta_{gen}$ – It is related to all electrical and mechanical losses in a wind generator, such as copper, iron, load, windage, friction, and other miscellaneous losses.

- Electric efficiency $\eta_{ele}$ – It encompasses all combined electric power losses in the converter, switches, controls, and cables.

Therefore, the total power conversion efficiency from wind to electricity $\eta_t$ is the production of these parameters, i.e.:

$$\eta_t = C_p \eta_{gear} \eta_{gen} \eta_{ele}$$

The effective power output from a wind turbine to feed into a grid becomes:

$$P_{eff} = C_p \eta_{gear} \eta_{gen} \eta_{ele} P_W = \eta_t P_w = \frac{1}{2}\left(\eta_t \rho A \bar{u}^3\right)$$

## Lanchester–betz Limit

The theoretical maximum efficiency of an ideal wind turbomachine was derived by Lanchester in 1915 and Betz in 1920. It was revealed that no wind turbomachines could convert more than 16/27 (59.26%) of the kinetic energy of wind into mechanical energy. This is known as Lanchester–Betz limit (or Lanchester– Betz law) today.

As shown in figure, $\bar{u}_1$ and $\bar{u}_4$ are mean velocities far upstream and downstream from the wind turbine; $\bar{u}_2$ and $\bar{u}_3$ are mean velocities just in front and back of the wind rotating blades, respectively. By assuming that there is no change in the air velocity right across the wind blades (i.e. $\bar{u}_2 = \bar{u}_3$) and the pressures far upstream and downstream from the wind turbine are equal to the static pressure of the undisturbed airflow (i.e. p1 = p4 = p ), it can be derived that:

$$p_2 - p_3 = \frac{1}{2}\rho\left(\bar{u}_1^2 - \bar{u}_4^2\right)$$

And,

$$\bar{u}_2 = \bar{u}_3 = \frac{1}{2}\left(\bar{u}_1 + \bar{u}_4\right)$$

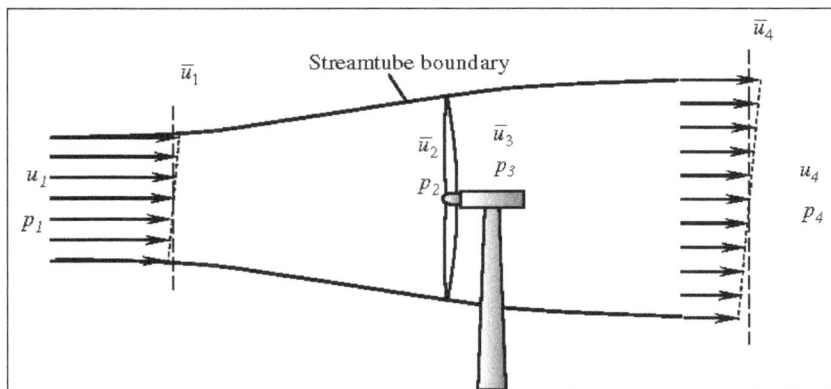

Airflow through a wind turbine.

Thus, the power output of mechanical energy captured by wind turbine blades is:

$$P_{me,out} = \frac{1}{2}\rho A \bar{u}_2\left(\bar{u}_1^2 - \bar{u}_4^2\right) = \frac{1}{2}\rho A \bar{u}_1^3 4a\left(1-a\right)^2$$

where a is the axial induction factor, defined as:

$$a = \frac{\overline{u}_1 - \overline{u}_2}{\overline{u}_1}$$

Substitute eqn $P_{\text{me,out}} = \frac{1}{2}\rho A \overline{u}_2 \left(\overline{u}_1^2 - \overline{u}_4^2\right) = \frac{1}{2}\rho A \overline{u}_1^3 4a\left(1-a\right)^2$ into $C_{\text{p}} = \frac{P_{\text{me,out}}}{P_{\text{w}}} = \frac{P_{\text{me,out}}}{\left(1/2\right)\rho A \overline{u}^3}$ (where $\overline{u}_1 = \overline{u}$), yields:

$$C_{\text{p}} = 4a\left(1-a\right)^2$$

This indicates that the power coefficient is only a function of the axial induction factor a. It is easy to derive that the maximum power coefficient reaches its maximum value of 16/27 when a = 1/3.

## Power Curve

As can be seen from eqn $P_{\text{eff}} = C_{\text{p}}\eta_{\text{gear}}\eta_{\text{gen}}\eta_{\text{ele}}P_{\text{W}} = \eta_{\text{t}}P_{\text{w}} = \frac{1}{2}\left(\eta_{\text{t}}\rho A \overline{u}^3\right)$, the effective electrical power output from a wind turbine Peff is directly proportional to the available wind power Pw and the total effective wind turbine efficiency $\eta_{\text{t}}$.

The power curve of a wind turbine displays the power output (either the real electrical power output or the percentage of the rated power) of the turbine as a function of the mean wind speed. Power curves are usually determined from the field measurements. As shown in figure, the wind turbine starts to produce usable power at a low wind speed, defined as the cut-in speed. The power output increases continuously with the increase of the wind speed until reaching a saturated point, to which the power output reaches its maximum value, defined as the rated power output. Correspondingly, the speed at this point is defined as the rated speed. At the rated speed, more increase in the wind speed will not increase the power output due to the activation of the power control. When the wind speed becomes too large to potentially damage the wind turbine, the wind turbine needs to shut down immediately to avoid damaging the wind turbine. This wind speed is defined as the cut-out speed. Thus, the cut-in and cut-out speeds have defined the operating limits of the wind turbine.

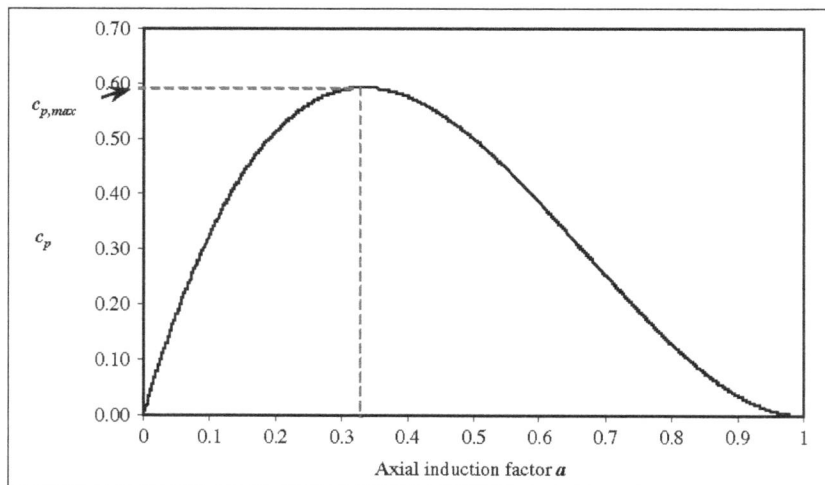

Power coefficient as a function of axial induction factor a.

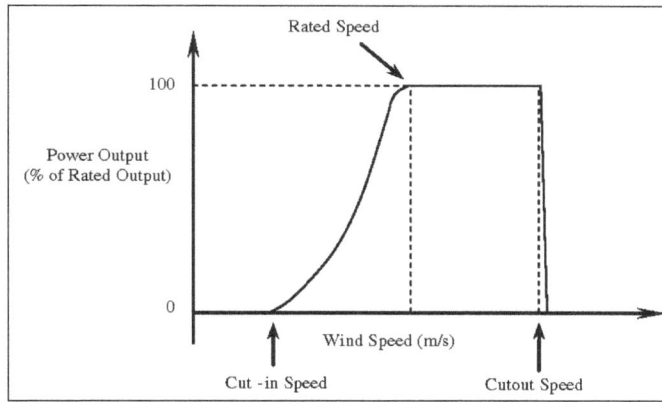

Typical wind turbine power curve.

There are a number of methods available for forecasting the wind turbine power performance curves. Based on statistical tools, a comparison of five different methods has been performed by Cabezon. The best results were obtained when the fuzzy logic tool and tuning over the transfer functions were applied for wind turbines. More recently, based on a stochastic model for the power conversion process, Gottschall and Peinke proposed a dynamic method for estimating the power performance curves and the dynamic approach has verified to be more accurate than the common IEC standard. A novel method, based on the stochastic differential equations of diffusive Markov processes, was developed to characterize wind turbine power performance directly from high-frequency fluctuating measurements.

## Tip Speed Ratio

The tip speed ratio is an extremely important factor in wind turbine design, which is defined as the ratio of the tangential speed at the blade tip to the actual wind speed, i.e.:

$$\lambda = \frac{(l+r)\omega}{\bar{u}}$$

where l is the length of the blade, r is the radius of the hub, and $\omega$ is the angular speed of blades.

If the blade angular speed w is too small, most of the wind may pass undisturbed though the blade swept area making little useful work on the blades. On the contrary, if $\omega$ is too large, the fast rotating blades may block the wind flow reducing the power extraction. Therefore, there exists an optimal angular speed at which the maximum power extraction is achieved. For a wind turbine with n blades, the optimal angular speed can be approximately determined as:

$$\omega_{opt} \approx \frac{2\pi}{n}\frac{\bar{u}}{L}$$

where L is the length of the strongly disturbed air stream upwind and downwind of the rotor. Substituting eqn $\omega_{opt} \approx \frac{2\pi}{n}\frac{\bar{u}}{L}$ into $\lambda = \frac{(l+r)\omega}{\bar{u}}$, the optimal tip speed ratio becomes:

$$\lambda_{opt} \approx \frac{2\pi}{n}\left(\frac{1+r}{L}\right)$$

Empirically, the ratio $(1 + r)/L$ is equal to about 2. Thus, for three-blade wind turbines (i.e. n = 3), $\lambda_{opt} \approx 4\pi/3$.

If the aerofoil blade is designed with care, the optimal tip speed ratio may be about 25–30% higher than the calculated optimal values above. Therefore, a wind turbine with three blades would have an optimal tip speed ratio:

$$\lambda_{opt} = \frac{4\pi}{3}(1.25 \sim 1.30) \approx 5.24 \sim 5.45$$

## Wind Turbine Capacity Factor

Due to the intermittent nature of wind, wind turbines do not make power all the time. Thus, a capacity factor of a wind turbine is used to provide a measure of the wind turbine's actual power output in a given period (e.g. a year) divided by its power output if the turbine has operated the entire time. A reasonable capacity factor would be 0.25–0.30 and a very good capacity factor would be around 0.40. In fact, wind turbine capacity factor is very sensitive to the average wind speed.

## Wind Turbine Controls

Wind turbine control systems continue to play important roles for ensuring wind turbine reliable and safe operation and to optimize wind energy capture. The main control systems in a modern wind turbine include pitch control, stall control (passive and active), yaw control, and others.

Under high wind speed conditions, the power output from a wind turbine may exceed its rated value. Thus, power control is required to control the power output within allowable fluctuations for avoiding turbine damage and stabilizing the power output. There are two primary control strategies in the power control: pitch control and stall control. The wind turbine power control system is used to control the power output within allowable fluctuations.

## Pitch Control

The pitch control system is a vital part of the modern wind turbine. This is because the pitch control system not only continually regulates the wind turbine's blade pitch angle to enhance the efficiency of wind energy conversion and power generation stability, but also serves as the security system in case of high wind speeds or emergency situations. It requires that even in the event of grid power failure, the rotor blades can be still driven into their feathered positions by using either the power of backup batteries or capacitors or mechanical energy storage devices.

Early techniques of active blade pitch control applied hydraulic actuators to control all blades together. However, these collective pitch control techniques could not completely satisfy all requirements of blade pitch angle regulation, especially for MW wind turbines with the increase in blade length and hub height. This is because wind is highly turbulent flow and the wind speed is proportional to the height from the ground. Therefore, each blade experiences different loads at different rotation positions. As a result, more superior individual blade pitch control techniques have been developed and implemented, allowing control of asymmetric aerodynamic loads on the blades, as well as structural loads in the non-rotating frame such as tower side-side bending. In such a control system, each blade is equipped with its own pitch actuator, sensors and controller.

In today's wind power industry, there are primarily two types of blade pitch control systems: hydraulic controlled and electric controlled systems. As shown in figure, the hydraulic pitch control system uses a hydraulic actuator to drive the blade rotating with respect to its axial centreline. The most significant advantages of hydraulic pitch control system include its large driving power, lack of a gear- box, and robust backup power. Due to these advantages, hydraulic pitch control systems historically dominate wind turbine control in Europe and North America for many years.

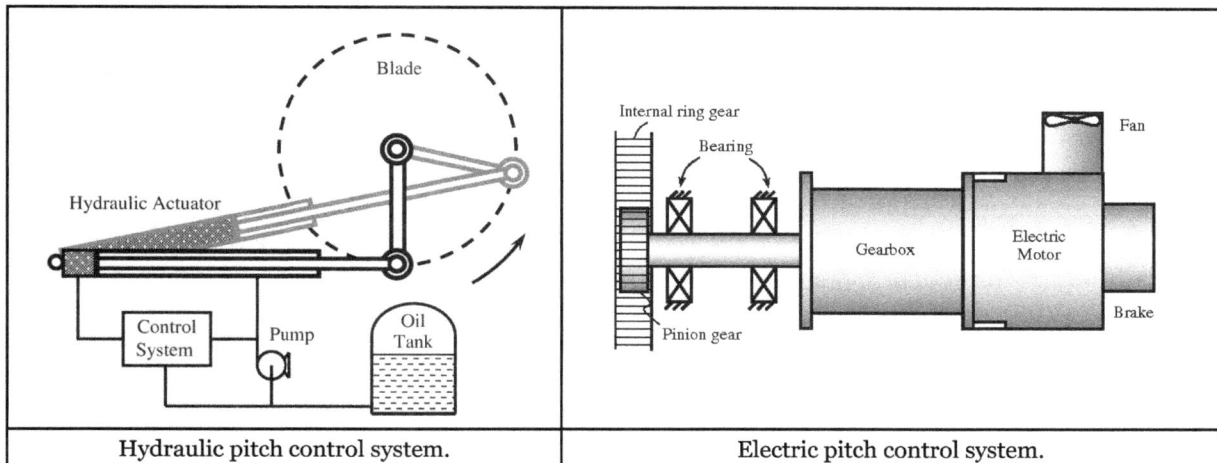

| Hydraulic pitch control system. | Electric pitch control system. |

The electric pitch control systems have been developed alternatively with the hydraulic systems. This type of control system has a higher efficiency than that of hydraulic controlled systems (which is usually less than 55%) and avoids the risk of environmental pollution due to hydraulic fluid being split or leaked.

In an electric pitch control system as shown in figure, the motor connects to a gearbox to lower the motor speed to a desired control speed. A drive pinion gear engages with an internal ring gear, which is rigidly attached to the roof of the rotor blade. Alternatively, some wind turbine manufacturers use the belt-drive structure adjusting the pitch angle. The use of electric motors can raise the responsiveness rate and sensitivity of blade pitch control. To enhance operation reliability, the use of redundant pitch control systems was proposed to be equipped in large wind turbines.

## Stall Control

Besides pitch control, stall control is another approach for controlling and protecting wind turbines. The concept of stall control is that the power is regulated through stalling the blades after rated speed is achieved.

Stall control can be further divided into passive and active control approaches. Passive stall control is basically used in wind turbines in which the blades are bolted to the hub at a fixed installing angle. In a passive stall-regulated wind turbine, the power regulation relies on the aerodynamic features of blades. In low and moderate wind speeds, the turbine operates near maximum efficiency. At high wind speeds, the turbine is automatically controlled by means of stalled blades to limit the rotational speed and power output, protecting the turbine from excessive wind speeds.

Compared with pitch control, a passive stall control system has a simple structure and avoids using a complex control system, leading to high reliability of the control system. In addition, the power fluctuations are lower for stall-regulated turbines. However, this control method has some disadvantages, such as lower efficiency, the requirement of external equipment at the turbine start, larger dynamic loads acting on the blades, nacelle, and tower, dependence on reliable brakes for the operation safety. Therefore, this control technique has been primarily used for small and medium wind turbines. Since the capacity of wind turbines has entered the multi-megawatt power range in recent years, pitch control has become dominant in the wind power market.

The active stall control technique has been developed for large wind turbines. An active stall wind turbine has stalling blades together with a blade pitch system. Since the blades at high wind speeds are turned towards stall, in the opposite direction as with pitch-control systems, this control method is also referred to as negative pitch control. Compared with passive stall control, active control provides more accurate control on the power output and maintains the rated power at high wind speeds. However, with the addition of the pitch-control mechanism, the active stall control mode increases the turbine cost and decreases operation reliability.

With megawatt wind turbines becoming the mainstream in the wind power industry from the late 1990s, pitch control is more favorable than stall control. It has been reported that the number of pitch-regulated turbines is four times higher than that of stall-regulated turbines and the trend is going to continue in coming decades.

## Yaw Control

In order to maximize the wind power output and minimize the asymmetric loads acting on the rotor blades and the tower, a horizontal-axis wind turbine must be oriented with rotor against the wind by using an active yaw control system. Like wind pitch systems, yaw systems can be driven either electrically or hydraulically. Generally, hydraulic yaw systems were used in the earlier time of the wind turbine development. In modern wind turbines, yaw control is done by electric motors. The yaw control system usually consists of an electrical motor with a speed reducing gearbox, a bull gear which is fixed to the tower, a wind vane to gain the information about wind direction, a yaw deck, and a brake to lock the turbine securely in yaw when the required position is reached. For a large wind turbine with high driving loads, the yaw control system may use two or more yaw motors to work together for driving a heavy nacelle.

In practice, the yaw error signals obtained from the wind vane are used to calculate the average yaw angle in a short interval. When this average yaw angle exceeds the preset threshold, the yaw motor is activated to align the turbine with the wind direction. Thus, with heavily filtered wind direction measurements, the actions of yaw control are rather limited and slow.

## Other Control Approaches

In the early time of wind turbine design, ailerons were once used to control the power output. This method involves placing moveable flaps on the trailing edge of rotor blades. The ailerons change the lift and drag characteristics of the blades and eventually change the rotor torque, which enable to regulate rotor speed and rotor power output. However, this method was less successful and was soon abandoned.

Another possibility is to yaw the rotor partly out of the wind to decrease power. This technique of yaw control is in practice used only for tiny wind turbines (>1 kW).

## Challenges in Wind Power Generation

While wind power generation offers numerous benefits and advantages over con- ventional power generation, there are also some challenges and problems need to be seriously addressed. The wide range of challenges and problems, from long- term environmental influences to thermal manage- ment of wind turbines, must be carefully considered in response to the rapid growth of wind power generation.

## Environmental Impacts

Modern wind farms today may contain a large number of large-size wind turbines. Therefore, their impacts on the environment cannot be ignored. One of the impacts is that poorly sited wind energy facilities may block bird migration routes and hurt or kill birds.

Though blade rotation speeds are rather low for large wind turbines at their normal operation, the tangential speeds at the blade tips could be higher than 70 m/s. At such high speeds, birds flying through the blade sweeping areas may be easily hurt or killed by colliding with blades. It has been reported by the US National Academy of Science that wind turbines may kill up to 40,000 birds per year in US. Though this number is much smaller than the 80 million birds killed by cars each year, it is important to evaluate the long-term influence on local geography, seasonal bird abundance and the species at risk. To reduce the bird death, using bird scares to drive birds away from wind farms has been considered. A more recent study has revealed that fossil-fuelled power stations appear to pose a much greater threat to avian wildlife than wind and nuclear power technologies.

Today, this problem becomes less important. Before building a wind farm, a series of environ- mental assessments have to be completed to avoid bird migration routes and to minimize other environmental impacts. Once the wind farm is built, further monitoring takes place to better un- derstand the ongoing relationship between birds and the wind farm.

Building wind farms will change the character of local landscape. Modern large wind turbines are more than 100 m tall and thus can be seen at a far distance. In practice, the visual effect for local residents is a significant consideration and is always scrutinized for wind projects. To minimize the visual effect, wind turbines usually use neutral colors such as light grey or off-white. Strate- gies to minimize visual effects involve the spacing, design, and uniformity of turbines, markings or lighting, roads and service buildings. There are a number of analytical tools available to assist understanding and testing of the effect of wind farms on visual amenity.

## Wind Turbine Noise

With the extensive build up of wind power plants and the population growth all over the world, the influence of wind turbine noise to the nearby residents becomes a problem not to be neglected. Wind turbine noise consists of aerodynamic noise from rotating blades and mechanical vibration noise from gearboxes and generators. For a modern large wind turbine, aerodynamic noise from the blades is considered to be the dominant noise source.

Though the noise limits vary significantly country to country, the approximate noise level at night-times in most European countries and Canada ranges from 35 to 40 dBA.

There are two components in aerodynamic noise: (1) airfoil self noise, that is, the noise produced by the blade in an undisturbed inflow and is caused by the interaction in the boundary layer with the blade trailing edge; and (2) inflow turbulence noise which is caused by the interaction of upstream atmospheric turbulence with the blade and depends on the atmospheric conditions. Both airfoil self noise and inflow turbulence noise mechanisms are dependent on a number of parameters such as wind speed, angle of attack, radiation direction, and airfoil shape.

There are a number of techniques for reducing aerodynamic noise produced by wind turbine blades. One of them is to use serrated blades at their trailing edges. It can improve blade aerodynamic characteristics and reduce the noise induced by Karman vortex street. Another is to use turbulence generating means, placed on the leeward surface side and at the outer section of the blade, to reduce noise. In a recent US patent application, it has reported that with an anti-noise device at the blade trailing edge, it allows altering the characteristics of the boundary layer and therefore modifies emitted noise.

The field measurements of GE wind turbines have shown that the use of the optimized blades and the serrated blades can reduce average overall noise by 0.5 and 3.2 dBA, respectively. In a field test of a 2.3 MW wind turbine, the over- all noise level reduction provided by blade serrations is over 6 dBA for at least two frequencies.

## Integration of Wind Power into Grid

Wind is a highly intermittent energy source for causing overall fluctuation in wind power generation. Electricity generated from wind turbines strongly depends on the local weather and geographic conditions that can fluctuate a great deal more than with some renewable energy sources such as hydropower.

With the increasing share of wind energy in the global power market, a large amount of wind power is integrated into existing grids. Thus, the expected growth in wind power could soon exceed the current capability of grids with today's technology. To prepare this situation in advance, the influence of intermittent wind power on the grid stability and system security must be properly addressed.

The impacts of wind power to a power grid depend on the level of wind power penetration, grid size and generation mix of electricity in the grid. Undoubtedly, there is no problem for low wind power penetration in a large power grid. However, integrating large utility-scale wind power presents unique challenges. These challenges call into questions such as: How to ensure system controllability? How to manage new kinds of variability and uncertainty. The detailed analysis regarding the impacts of wind power on power systems can be found in.

## Thermal Management of Wind Turbines

Large wind turbines are usually installed far away from urban areas and often operate under severe climate conditions, thus experiencing large variations in environmental temperatures. As a consequence, there is a need for a wind turbine to have a robust thermal control system for maintaining temperature levels inside the nacelle within specified limits.

During turbine operation, heat is generated from electric/electronic devices and rotating mechanical components (e.g. gearboxes and bearings) as a result of various power losses. For ensuring safe and reliable operation and preventing failure of the turbine, heat generated in the wind turbine must be dissipated efficiently.

Wind turbine cooling includes:

- Wind generator cooling.

- Electronic and electric equipment cooling.

- Gearbox cooling.

- Other components/subsystems cooling.

New cooling techniques have continuously been innovated in all cooling modes. A method was proposed to utilize incoming wind to cool the wind turbine. This wind assisted cooling system sucks in wind flow from an air inlet port on the top of the nacelle, fills the received airflow into the generator and finally exhausts at the front of the nacelle. Some large wind generators use water or oil cooling for dealing with high thermal loads. While the turbine benefits high cooling efficiency, it also suffers lower reliability and higher cost for adding such a complex cooling system.

The main challenge for electronic devices in a wind turbine is that they must withstand a wide range of ambient temperatures, usually from −40 to +55 °C. In addition, they must be protected from dusts and moisture, as well as electrical shocks from lightning. There are several cooling modes in electronic cooling, including passive or active air cooling, forced single- or multi-phase liquid cooling, and phase change cooling. Under high ambient temperature conditions, a cooling or ventilation system is necessary to prevent overheating of electronic devices.

In cold climates, heating may be required for:

- Warming up the lubrication oil in gearboxes.

- Heating blades and hub to prevent them from icing over.

- Raising the temperature inside the control cabinets toward a desired temperature range to prevent electronic devices from malfunctioning.

## Wind Energy Storage

Today developing advanced, cost-effective storage technologies of electric energy still remains a challenge, which may limit the widespread application of wind energy. The research and development (R&D) of new energy storage systems are highly desired to meet cyclical energy demands and stabilize power output, especially for large-scale wind farms.

The technologies for wind energy storage have been developed over several decades to convert wind energy into various forms of energy, including:

- Electrochemical energy in batteries and super capacitors.

- Magnetic energy in superconducting magnetic energy storage (SMES).

- Kinetic energy in rotating flywheels.

- Potential energy in pumped water at higher altitudes.

- Mechanical energy in compressed air in vast geologic vaults.

- Hydrogen energy by decomposing water.

Among these techniques, the most popular method is to use batteries. However, there are some drawbacks to regular batteries, such as cost, short lifetime, corrosion, and disposal concerns. Research and development of innovative batteries are underway. It has reported that lithium-ion battery technology is projected to provide stationary electrical energy solutions to enable the effective use in renewable energy sources. It is expected that safe and reliable lithium-ion batteries will soon be connected to solar cells and wind turbines. Sodium-sulfur battery is another promising candidate for energy storage. This type of batteries is preferably used to store renewable energy such as wind, sunlight, and geothermal heat.

## Wind Turbine Lifetime

Modern wind turbines are designed for the lifetime of 20–30 years. A critical challenge facing turbine manufacturers and wind power plants is how to achieve the lifetime goals while at the same time minimize the costs of maintenance and repair. However, improving the operational reliability and extending the lifetime of wind turbines are very difficult tasks for a number of reasons:

- Wind turbines have to be exposed to various hostile conditions such as extreme temperatures, wind speed fluctuations, humidity, dust, solar radiation, lightning, salinity and frequent onslaughts of rain, hail, snow, ice, and sandstorms.

- A modern wind turbine consists of a large number of components and systems; each of them has its own lifetime. According to the Cannikin law, failure must first occur in the component or system with the shortest lifetime.

- A wind turbine is subjected to a large variety of dynamic loads due to wind fluctuations in speed and direction and numerous starts and stops of the system. Some primary parts or components have to withstand heavy fatigue loads.

- Advanced high-strength, fatigue-resistant materials are vital to some key components in modern large wind turbines due to the continuous increase in blade length, hub height, and turbine weight.

- As a complex engineering system, a wind turbine must be designed at the system level rather than part/component level as a common practice in some turbine manufacturers.

## Cost of Electricity from Wind Power

- Although the wind power industry appears to be booming in recent years world- wide, achieving continuous cost reduction in wind power generation continues to be a challenge and a key focus for the wind industry.

- Wind power is characterized by low variable costs and relatively high fixed costs. The main factors governing wind power economics are: Investment costs, including wind turbines, foundations, and grid connection.

- Operation and maintenance (O&M) costs, including regular maintenance, repairs, insurance, spare parts, and administration.

- Wind turbine's electricity production cost, which highly depends on the wind turbine capacity, wind farm size, and average wind speed at the chosen site.

- Wind turbine lifetime.

- Discount rate.

- Among these, the most important factors are the wind turbines' electricity production and their investment costs. The trends towards lager wind turbines and larger wind farms help reduce both investment and O&M costs per kilowatt-hour (kWh) produced.

- Though the price of electricity from wind has fallen approximately 90% over the last 30 years because of the developments of wind technology, it is still more expensive than those from coal or natural gas. It has been predicted by Electric Power Research Institute that even for plants coming online in 2015, wind energy would cost nearly one-third more than coal and about 14% more than natural gas. This is the greatest obstacle for wind power to increase its share in the electric power market. A recent study indicates that wind energy in US today still depends on federal tax incentives to compete with fossil fuel prices, and technology progress could dominate future cost competitiveness.

- The global financial and economic crisis, which started from early 2008, has dramatically altered the pace of wind development. With reduced power consumption, the prices of fossil fuels (e.g. coal and natural gas) have greatly decreased, putting even more pressures on the wind power industry to continuously drive down wind power costs for staying competitive in the present challenging economic times.

## Trends in Wind Turbine Developments and Wind Power Generation

Wind turbine technology has been developed by continuously optimizing turbine design, improving turbine performance, and enhancing overall turbine efficiency. There have been several generations of development and improvement in wind turbine technology, concentrated on blades, generators, direct drive techniques, pitch and yaw control systems, and so on. To provide more electrical energy from wind technology in the next several decades, it requires:

- Developing innovative techniques.

- Decreasing wind turbine costs through technology advancement.

- Optimizing manufacturing processes and enhancing manufacturing operations.

- Improving wind turbine performance and efficiency.

- Reducing operating and maintenance costs.

- Expanding wind turbine production capacities. The current major trends in the development of wind turbines are towards higher power, higher efficiency and reliability, and lower cost per kilowatt machines.

## High-power and Large-capacity Wind Turbine

One of the significant developments in wind turbine designing and manufacturing in recent years is the increase in the wind turbine capacity of individual wind turbines. From machines of just 25 kW two decades ago, the commercial range of modern wind turbines sold today is typically 1–6 MW. At the same time, 7–10 MW wind turbines are underway in some larger wind turbine OEMs. With this trend, innovative techniques have been developed and new materials have been adopted for optimizing the wind turbine performance and minimizing the operation and manufacturing costs. Enercon has installed the present world's largest wind turbine E-126 in Germany and is in the process of installing more units in Belgium. The E-126 turbine is rated at 6 MW with the rotor diameter of 126 m. Clipper Windpower has announced that it is planning to build a 7.5 MW offshore wind turbine.

However, while high-powered wind turbines enable to increase wind power output per unit and lower the cost per kWh, there are some significant challenges facing wind turbine engineers:

- Failure rates of wind turbines depend not only on turbines' operational age but also their rated power. High-power, large-size wind turbines have shown significant higher annual failure rates due to the primary failures of the control system, drivetrain, and electronic/electrical components. Because most of mega-watt wind turbines were usually among the first models installed, they show high early failure rates that decrease slightly throughout their years of operation.

- Wind velocity is proportional to the height from the earth's surface. With the continuously increasing blade length of large wind turbines, the differences of the dynamic wind loads between the rotating blades become significantly large, resulting in a large resultant unbalanced fatigue load on the turbine blades, and a resultant unbalanced torsional moment on the main shaft, and in turn, on the wind tower.

- During wind turbine's operation, a minimum clearance must be maintained between the blade tips and the wind tower. Therefore, high blade stiffness is required to avoid the collision between the blades and the tower. In practice, the maximum blade length is constrained by required stiffness and stresses of blades.

- Large wind turbines become more susceptible to variations in wind speed and intensity across the swept area.

- Transportation and installation of long-length blades remain challenges to the wind power industry. The length of a blade for a 4.5–5 MW wind turbine ranges 50–70 m. It is very difficult to ship such long blades through current highways and installed on the top of 120–160 m wind towers.

- The tower strength is another consideration. For a given survivable wind speed, the mass of a wind turbine is approximately proportional to the cube of its blade length and the output power is proportional to the square of it blade length. Typically, the mass of a 4.5–5 MW is of 200–500 tons. It was reported that doubling the tower height generally requires doubling the diameter as well, increasing the amount of material by a factor of 8.

To ensure the sustainability of the increase in power output and turbine size, all these challenges must be carefully and effectively addressed.

## Offshore Wind Turbine

With several decades of experience with onshore wind technology, offshore wind technology has presently become the focus of the wind power industry. Due to the lower resistance, wind speeds over offshore sea level are typically 20% higher than those over nearby lands. Thus, according to the wind power law, the offshore wind power can capture much more power than the onshore one. This indicates that an offshore wind turbine may gain a higher capacity factor than that of its land-based counterpart. In addition, because the offshore wind speeds are relatively uniform with the lower variations and turbulence, it enables the offshore wind turbines to simplify the control systems and reduces blade and turbine wears.

Sweden installed the first offshore wind turbine in 1990, with the unit capacity of 220 kW. Denmark built its first demonstration offshore wind turbines in 1991, which consists of 11 units, with the unit capacity of 450 kW. With the developments of offshore wind technology in the next several years, offshore wind turbines entered the stage of industrial production in 2001. Today, high capacity wind turbines focus on the offshore application. In 2009, nearly 600 MW offshore wind power were added and connected to electric grids, basically by European countries, bringing the total accumulative installed offshore wind power capacity to more than 2,000 MW. It is expected that in 2010 ten additional European offshore wind farms will be completed to add approximately 1,000 MW online, which represents a growth rate of 75% compared to 2009. The related foundation technologies for offshore applications are also being developed for the erection of higher capacity wind turbines. The annual installed offshore wind power capacities from 1990 to 2009 are shown in figure.

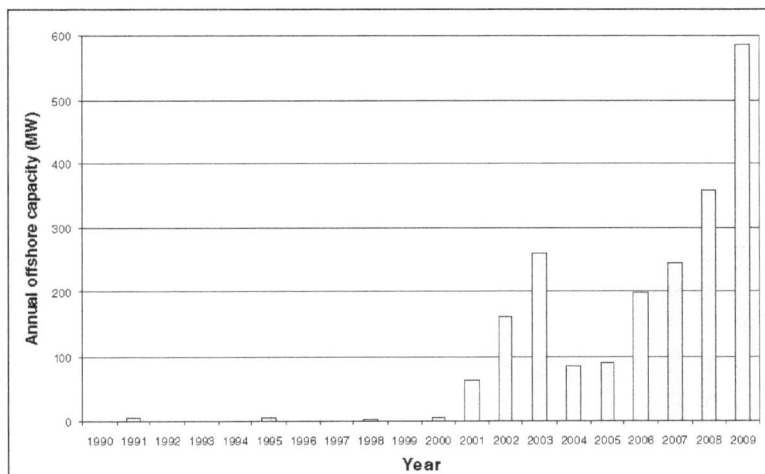

Annual installed offshore wind power capacity.

Offshore wind turbines are installed in seawater, with greater risks of structural corrosion, particularly under conditions of high wave, sea salt splashing and low temperature. To avoid or at least delay the corrosion to protect wind turbines, a number of techniques have been developed.

One of them involves the use of electrochemical reactions to prevent the steel corrosion, known as cathodic protection. This is done using a small negative voltage applied to metal. A new method, called impressed current cathodic protection, was invented by Brown and Hefner. This method includes providing an impressed current anode electrochemically coupled to the wind turbine support structure to the impressed current anode to operate the impressed current anode.

In dealing with corrosion control, coating technology plays an important role in the wind power industry. The adoption of high performance of multi-coating systems has achieved satisfactory results to prevent external and internal corrosions of wind turbines. The coating materials used today include thermally sprayed metal (zinc and aluminum), siloxane, acrylic, epoxy, polyurethane, etc. It is expected to develop inorganic hybrid materials (such as polysiloxane) as super durable finishes to meet more aggressive environments.

## Direct Drive Wind Turbine

In a direct drive wind turbine, rotor blades directly drive the rotor of the wind generator. By eliminating the multi-stage gearbox, which is one of the most easily damaged components in a MW wind turbine, the number of rotating parts is greatly reduced and the turbine structure is considerably simplified. As a result, it significantly increases the reliability and efficiency of the wind turbine and reduces turbine noise and maintenance costs. Since direct drive wind generators operate at relative low speeds, it reduces the wear and tear of the generator. In order to identify suitable generator concepts for director drive wind turbines, Bang have compared various direct drive generator systems and concluded that direct drive permanent-magnet synchronous machines are more superior in terms of the energy yield, reliability and maintenance costs.

Though the concept of direct drive wind turbine has been proposed for a long time, the modern direct drive techniques have become available until recent three decades. Presently, direct drive wind turbines have been manufactured by large wind OEMs. Siemens installed two innovative 3.6 MW direct drive wind turbines at a site in west Denmark in 2008. The feasibility of building 10 MW direct-drive wind turbine was investigated by Polinder.

However, direct drive wind turbines have some disadvantages in terms of the cost, size, and mass, making them difficult in manufacturing, shipping, and installing. Without a gearbox, the rotor diameter in a direct drive generator must be made larger enough to maintain a relative high rotating speed at the air gap. A lot of structural material must be added to keep the stator and rotor in place for maintaining the air gap. Therefore, the direct drive wind turbine has a larger size and a higher weight. For instance, the Siemens 3.6 MW direct drive wind turbine has a total weight of 265 tons (nacelle 165 tons and rotor 100 tons), as compared with 235 tons for a 3.6 MW geared turbine. This requires the higher strength for the turbine tower. According to Bang, the cost of a 3 MW direct drive PM synchronous generator system could be 35% higher than that of a 3 MW induction genera- tor system with three stage gearbox. To make the direct drive wind turbines more attractive to the wind market, all these disadvantages must be solved.

## High Efficient Blade

Rotor blade design can be split into structural and aerodynamic design. During normal wind turbine operation, rotor blades have to withstand enormous dynamic loads. The bending moment due to the gravity load results in up to $10^8$ load cycle alternations within the turbine lifetime. In addition, there are stochastic alternating loads caused by wind turbulence and the effects of ageing of the materials due to the weather. As wind turbines become larger and larger, the length, size, and weight of blades increase accordingly. For instance, the blade diameter in a large wind turbine could be longer than 100 m, which is higher than the wingspan and length of Boeing 747-400 at 64.4 and 70.7 m, respectively. There is no doubt that these blades require extremely high fatigue strength.

In the blade structural design, one indicator is the blade weight/swept area ratio (or swept area density in some references). It is highly desired to minimize this ratio while satisfying the blade strength requirements. Most blades used today are made from composite materials such as glass-fibre epoxy, carbon epoxy, fibre-reinforced plastic, etc. With epoxy resin/glass-fibre material, the weight/swept area ratio of 1–1.5 kg/m2 can be achieved up to a rotor diameter of 62 m. With the trend toward long-length and larger-size blades, high- strength, fatigue-resistant materials such as metallic materials need to be considered.

The aerodynamic design of wind turbine blades is important as it determines the wind energy capture. With advanced CFD tools, the shape of aerodynamic profile and dimensions of blade can be preliminarily determined and the blade optimization can be achieved via field tests. As a good example, a study of aerodynamic and structural design for wind turbines larger than 5 MW was reported by Hillmer.

There are a number of improvements achieved in the rotor blade design. A new type of wind turbine blades, named "STAR" (Sweep Twist Adaptive Rotor) blades, was specially designed for low-wind-speed regions. The test results have shown that the STAR blades can shed 20% of the root moment via tip twist of about 3° and yield 5–10% annual energy capture than the regular blades.

Researchers have developed an innovative technique that uses sensors and computational software to constantly monitor forces exerted on wind turbine blades. The data is fed into an active control system that precisely adjusts the shape of rotor blades to respond to changing winds. The technique could also help improve turbine reliability by providing critical real-time information to the control system for preventing damage to blades from high winds. Recent aerodynamic research has revealed that an increase of the aerodynamic efficiency of a wind turbine rotor may be achieved by extending the turbine blades to very close to the wind turbine nacelle.

## Floating Wind Turbine

Dr. Sclavounos at MIT is among the first to develop the concept of floating wind turbines in deep water. He and his team in 2004 integrated a wind turbine with a floater. According to their analysis, the floater-mounted turbines could work in water depths of up to 200 m.

The world's first commercial-scale floating wind turbine has being constructed in deep water far from land. The turbine is mounted on a floating turbine platform on the sea surface and anchored to the seabed with three strong chains. Changing the length of the chains could allow the turbine to operate in water depths between 50 and 300 m, enough to take it far out into the deep ocean. By comparing with existing offshore wind turbines, floating wind turbine is more economic in the installation and shipping. Electricity would be sent ashore using undersea cables.

Sway, a Norwegian company, plans to launch its prototype of floating wind turbines in 2010. The turbine is to be mounted on an elongated floating mast, connected to the seabed by a metal tube. The turbine mast is designed to sway with wind and waves, and can lean at an angle of up to 15°.

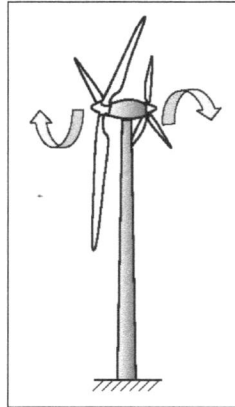

A dual-blade set wind turbine.

## Wind Turbine with Contra-rotating Rotors

Coaxial contra-rotating propellers have been widely applied to aircrafts, marines, ships, and torpedoes for improving the propulsion efficiency and offsetting system reactive torques. The efficiency of contra-rotating propellers was found 6–16% higher than that of normal propellers. Late, the contra-rotating concept has been introduced into the wind turbine design. The contra-rotating type of wind turbines has usually two sets of blades that rotate in opposite directions. These two sets of blades can be arranged either one behind another at the turbine front, or separately at the turbine front and rear.

Schönball is one of the first to describe a mechanism which composes of two contra-rotating rotors. In the generator, a rotor is driven by one wind wheel and a regular stationary stator driven by another wind wheel. Owing to the opposed rotation of two wind wheels, the relative speed between the two rotors may be doubled and thus the efficiency is correspondingly increased, compared with a conventional generator having one rotor. Similarly, McCombs developed a wind turbine equipped with two sets of blades that propel the rotor and stator directly. More recently, Wachinski proposed a drive device for an improved windmill composed of two counter-rotative sets of blades. In 2005, with two granted US patents, Kowintec Inc. successfully built up a 100 kW and 1 MW contra-rotating wind turbine prototypes. The test results have shown that with the dual-blade and dual-rotor, the power production efficiency of the turbine increases 24% and the system cost decreases 20–30%, by comparing with a conventional single rotor unit.

he benefits of using contra-rotating wind turbines are:

- Enhancing wind energy capturing capability – A wind turbine with a contrarotating system can capture more wind energy to convert into electricity, by comparing with a single set of blades. In a contra-rotating win turbine, each set of blades contributes independently to the total power output. Because the length of each set of blades can be made differently, the wind turbine may produce electricity at a lower cut-in speed by the set of blades with a shorter blade length.

- Achieving higher power density – With the increase in power output and the moderate increase in wind turbine volume, the power density becomes higher.

- Reducing wind turbine cost per kWh – Though the total cost of this type of wind turbines increases due to the addition of extra parts, the cost per kWh will decrease for the large increase in power output.

- Increasing wind turbine operation reliability – Some types of contra-rotating wind turbines eliminate gearboxes to simplify the design and increase the turbine operation reliability.

However, this type of wind turbine also has some drawbacks. The main negative effect is the wake vortices created by the first set of blades which can substantially lower the performance of the second set of blades. In fact, the wake vortices will enhance the wind turbulence intensity to strengthen unbalanced dynamic loads on the second set of blades, decreasing the mean wind velocity and the power captured by the second set of blades.

## Drivetrain

In a geared wind turbine, the term "drivetrain" usually encompasses all rotating parts, from the rotor hub, the main gearbox, to the generator. The main gearbox is one of the most important components in the wind turbine, for increasing the slow rotation speed of the blades to the desired high speed of the generator's rotor. It is also the most expensive component in the turbine and can easily fail before reaching the intended life. With the increase in the turbine size and capacity in the last decade, the gearbox has been subjected to even greater loads and stresses. There are significant challenges presented to gearbox designers and manufacturers.

Wind turbines typically use planetary gears to divide torque along three paths and reduce individual loads on each gear. However, torsional loads twist gears out of alignment, and slight dimensional variation in gearbox components, indicate that planetary gears do not equally share the load. Misaligned gears, shock loads, and uneven forces lead to highly localized stress and eventually fracture along the gear edges. To solve these problems, an innovative type of gearbox has been developed, using the Integrated Flex-pin Bearing (IFB) to equalize gear loads, eliminate misalignment, and dramatically improve wind turbine reliability. This novel design increases torque capacity of planetary gears up to 50%.

Based on the failure analysis of gearboxes, the Gearbox Reliability Collaborative initiated at US National Renewable Energy Laboratory (NREL) provides a fresh approach toward better gearboxes that combines the resources of key members of the supply chain to investigate design-level root causes of field problems and solutions that will lead to higher gearbox reliability.

Flex-drive planetary gearbox.

## Integration of Wind and other Energy Sources

One of the notable characteristics in the wind power generation is its uncertainty due to the sudden change in both wind speed and direction, especially for off-grid wind power generation systems. Therefore, the power output from wind turbines fluctuates from time to time. When wind turbines are connected to a small or isolated grid, the power output from other generators must be varied in response to these variations and fluctuation in order to keep system frequency and voltage within predefined limits. For this purpose, it is beneficial to integrate wind and other complementary energy sources to form hybrid power systems for assuring the stability and reliability of power supply and reducing the requirement for the wind energy storage.

## Wind–solar Hybrid System

Both wind and solar energy are highly intermittent electricity generation sources. Time intervals within which fluctuations occur span multiple temporal scales, from seconds to years. These fluctuations can be subdivided into periodic fluctuations (diurnal or annual fluctuations) and non-periodic fluctuations related to the weather change.

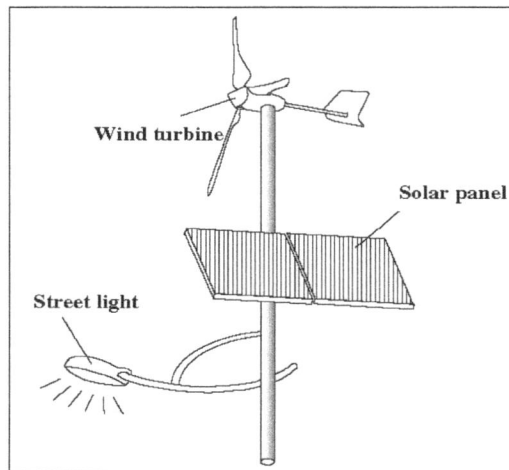

Wind–solar hybrid system for street lights.

Wind and solar energy are complementary to each other in time sequence and regions. In the summer, sunlight is intensive and the sunshine duration is long but there is less wind. In the winter, when less sunlight is available, wind becomes strong. During a day, the sunshine is strong while wind is weak. After sunset, the wind is strengthened due to large temperature changes near the earth's surface. It has been reported that the effects of complementarity are more dramatic in certain periods and locations at Serbia. The analyses and test data of wind–solar hybrid power systems have shown that the optimum combination of the wind– solar hybrid system lies between 0.70 and 0.75 of solar energy to load ratio with the minimized life cycle cost. For all load demands, the leveled energy cost for the wind–solar hybrid system is always lower than that of standalone solar or wind system.

Because the major operating time for wind and solar systems occurs at different periods of time, wind–solar hybrid power systems can ensure the reliability of electricity supply. The applications of wind–solar hybrid systems ranges extensively from residential houses to municipal and industrial facilities, either grid- connected or standalone. For instance, as an independent power supply source, wind–solar hybrid systems have been widely used in China for street lighting. The world's largest wind–solar power test base, integrating wind power, photovoltaic power and energy storage, is being constructed at Zhangbei, China. The project will have an installed capacity to generate 300 MW of wind power, 100 MW of solar power and 75 MW of chemical energy storage.

## Wind–hydro Hybrid System

Hydropower generation is to convert potential energy in water into electrical energy by means of hydropower generators. As a renewable and clean energy source, hydropower accounts for the dominant portion of electricity generated from all renewable sources.

In many locations of the world, hydropower is complementary with wind power, while the seasonal wind power distribution is higher in winter and spring but lower in summer and fall, hydropower is lower in the dry seasons (winter and spring) but higher in the wet seasons (summer and fall). Thus, the integration of wind and hydropower systems can provide significant technical, economic, and systematic benefits for both systems. Taking a reservoir as a means of energy regulation, "green" electricity can be produced with wind–hydro hybrid systems.

## Wind–hydrogen System

Hydrogen is an energy carrier and can be produced from a variety of resources such as water, fossil fuels, and biomass. As a fuel with a high energy density, hydrogen can be stored, transported and then converted into electricity by means of fuel cells at end users. It is widely recognized that wind power, solar power and other renewable energy power generation systems can be integrated with the electrolysis hydrogen production system to produce hydrogen fuel. The largest wind-to-hydrogen power system in the UK has been applied to a building that is fuelled solely by wind and "green" hydrogen power with the developed hydrogen mini-grid system technology. In this system, electricity generated from a wind turbine is mainly used to provide to the building and excess electricity is used to produce hydrogen using a state-of-the-art high-pressure alkaline electrolyser.

## Wind–diesel Power Generation System

Wind power can be combined with power produced by diesel engine-generator systems to provide a stable supply of electricity. In response to the variations in wind power generation and electricity consumption, diesel generator sets may operate intermittently to reduce the consumption of the fuel. It was reported that a viable wind–diesel stand-alone system can operate with an estimated 50–80% fuel saving compared to power supply from diesel generation alone.

Wind–diesel hybrid power systems have been studied since 1995 in the US. Till now, many new techniques have been developed and a large number of wind– diesel power generation systems have been installed all over the world. According to the proportion of wind use in the system, three different types of wind–diesel systems can be distinguished: low, medium, and high penetration wind–diesel systems. Presently, low penetration systems are used at the commercial level, whereas solutions for high penetration wind–diesel systems are at the demonstration level. The technology trends include the development of robust and proven control strategies.

# Wind Power

Wind power stations.

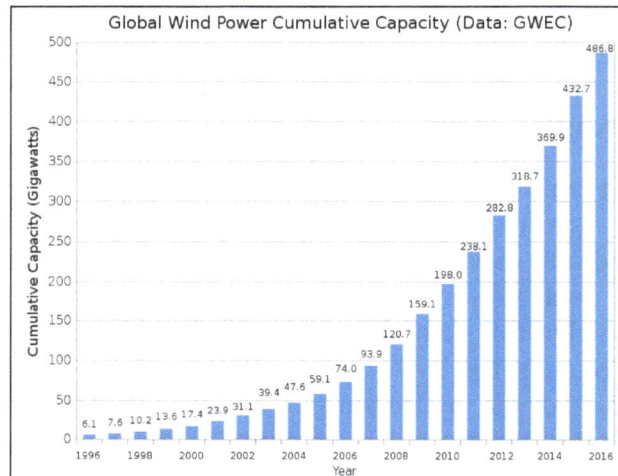

Global Growth of Installed Capacity.

Wind power or wind energy is the use of wind to provide the mechanical power through wind turbines to turn electric generators and traditionally to do other work, like milling or pumping. Wind power is a sustainable and renewable energy, and has a much smaller impact on the environment compared to burning fossil fuels.

Wind farms consist of many individual wind turbines, which are connected to the electric power transmission network. Onshore wind is an inexpensive source of electric power, competitive with or in many places cheaper than coal or gas plants. Onshore wind farms also have an impact on the landscape, as typically they need to be spread over more land than other power stations and need to be built in wild and rural areas, which can lead to "industrialization of the countryside" and habitat loss. Offshore wind is steadier and stronger than on land and offshore farms have less visual impact, but construction and maintenance costs are higher. Small onshore wind farms can feed some energy into the grid or provide electric power to isolated off-grid locations.

Wind is an intermittent energy source, which cannot make electricity nor be dispatched on demand. It also gives variable power, which is consistent from year to year but varies greatly over shorter time scales. Therefore, it must be used together with other electric power sources or storage to give a reliable supply. As the proportion of wind power in a region increases, more conventional power sources are needed to back it up (such as fossil fuel power and nuclear power), and the grid may need to be upgraded. Power-management techniques such as having dispatchable power sources, enough hydroelectric power, excess capacity, geographically distributed turbines, exporting and importing power to neighboring areas, energy storage, or reducing demand when wind production is low, can in many cases overcome these problems. Weather forecasting permits the electric-power network to be readied for the predictable variations in production that occur.

In 2018, global wind power capacity grew 9.6% to 591 GW. In 2017, yearly wind energy production grew 17%, reaching 4.4% of worldwide electric power usage, and providing 11.6% of the electricity in the European Union. Denmark is the country with the highest penetration of wind power, with 43.4% of its consumed electricity from wind in 2017. At least 83 other countries are using wind power to supply their electric power grids.

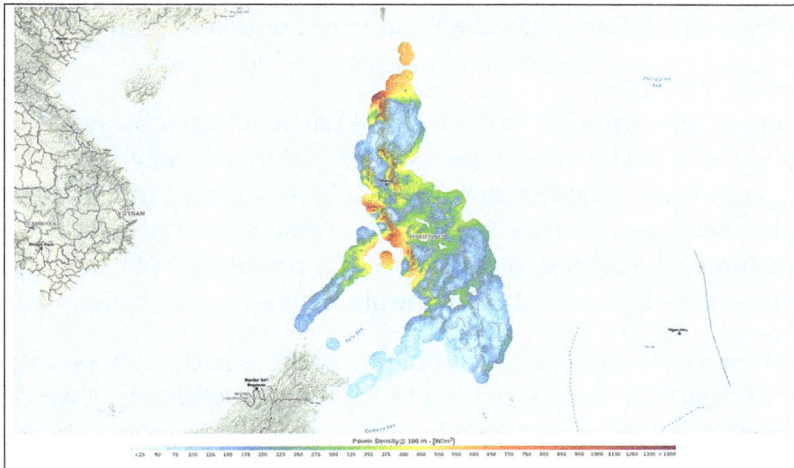

Philippines wind power density map at 100 m above surface level.

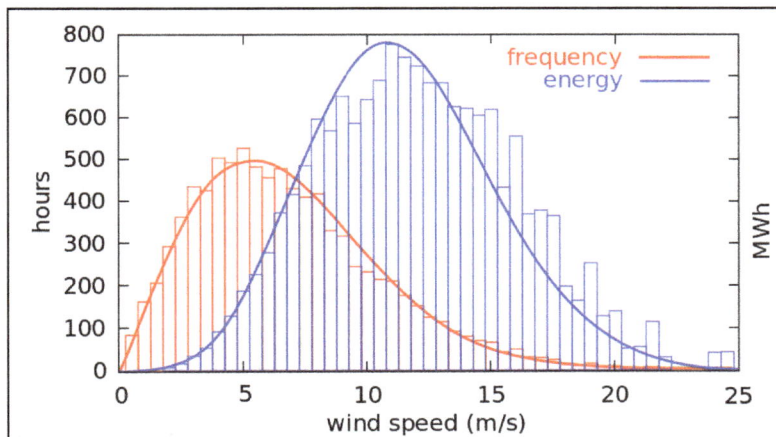

Distribution of wind speed (red) and energy (blue) for all of 2002 at the Lee Ranch facility in Colorado. The histogram shows measured data, while the curve is the Rayleigh model distribution for the same average wind speed.

Wind energy is the kinetic energy of air in motion, also called wind. Total wind energy flowing through an imaginary surface with area A during the time t is:

$$E = \frac{1}{2}mv^2 = \frac{1}{2}(Avt\rho)v^2 = \frac{1}{2}At\rho v^3$$

where ρ is the density of air; v is the wind speed; Avt is the volume of air passing through A (which is considered perpendicular to the direction of the wind); Avtρ is therefore the mass m passing through "A". ½ ρv² is the kinetic energy of the moving air per unit volume.

Power is energy per unit time, so the wind power incident on A (e.g. equal to the rotor area of a wind turbine) is:

$$P = \frac{E}{t} = \frac{1}{2}A\rho v^3.$$

Wind power in an open air stream is thus proportional to the third power of the wind speed; the available power increases eightfold when the wind speed doubles. Wind turbines for grid electric power therefore need to be especially efficient at greater wind speeds.

Wind is the movement of air across the surface of the Earth, affected by areas of high pressure and of low pressure. The global wind kinetic energy averaged approximately 1.50 MJ/m² over the period from 1979 to 2010, 1.31 MJ/m² in the Northern Hemisphere with 1.70 MJ/m² in the Southern Hemisphere. The atmosphere acts as a thermal engine, absorbing heat at higher temperatures, releasing heat at lower temperatures. The process is responsible for production of wind kinetic energy at a rate of 2.46 W/m² sustaining thus the circulation of the atmosphere against frictional dissipation.

Through wind resource assessment it is possible to provide estimates of wind power potential globally, by country or region, or for a specific site. A global assessment of wind power potential is available via the Global Wind Atlas provided by the Technical University of Denmark in partnership with the World Bank. Unlike 'static' wind resource atlases which average estimates of wind speed and power density across multiple years, tools such as Renewables.ninja provide time-varying simulations of wind speed and power output from different wind turbine models at an hourly resolution. More detailed, site specific assessments of wind resource potential can be obtained from specialist commercial providers, and many of the larger wind developers will maintain in-house modeling capabilities.

The total amount of economically extractable power available from the wind is considerably more than present human power use from all sources. Axel Kleidon of the Max Planck Institute in Germany, carried out a "top down" calculation on how much wind energy there is, starting with the incoming solar radiation that drives the winds by creating temperature differences in the atmosphere. He concluded that somewhere between 18 TW and 68 TW could be extracted.

Cristina Archer and Mark Z. Jacobson presented a "bottom-up" estimate, which unlike Kleidon's are based on actual measurements of wind speeds, and found that there is 1700 TW of wind power at an altitude of 100 metres over land and sea. Of this, "between 72 and 170 TW could be extracted in a practical and cost-competitive manner". They later estimated 80 TW. However research at Harvard University estimates 1 watt/m² on average and 2–10 MW/km² capacity for large scale wind farms, suggesting that these estimates of total global wind resources are too high by a factor of about 4.

The strength of wind varies, and an average value for a given location does not alone indicate the amount of energy a wind turbine could produce there.

To assess prospective wind power sites a probability distribution function is often fit to the observed wind speed data. Different locations will have different wind speed distributions. The Weibull model closely mirrors the actual distribution of hourly/ten-minute wind speeds at many locations. The Weibull factor is often close to 2 and therefore a Rayleigh distribution can be used as a less accurate, but simpler model.

## Wind Farms

A wind farm is a group of wind turbines in the same location used for production of electric power. A large wind farm may consist of several hundred individual wind turbines distributed over an extended area, but the land between the turbines may be used for agricultural or other purposes. For example, Gansu Wind Farm, the largest wind farm in the world, has several thousand turbines. A wind farm may also be located offshore.

Almost all large wind turbines have the same design — a horizontal axis wind turbine having an upwind rotor with three blades, attached to a nacelle on top of a tall tubular tower.

In a wind farm, individual turbines are interconnected with a medium voltage (often 34.5 kV) power collection system and communications network. In general, a distance of 7D (seven times the rotor diameter of the wind turbine) is set between each turbine in a fully developed wind farm. At a substation, this medium-voltage electric current is increased in voltage with a transformer for connection to the high voltage electric power transmission system.

## Generator Characteristics and Stability

Induction generators, which were often used for wind power projects in the 1980s and 1990s, require reactive power for excitation, so substations used in wind-power collection systems include substantial capacitor banks for power factor correction. Different types of wind turbine generators behave differently during transmission grid disturbances, so extensive modelling of the dynamic electromechanical characteristics of a new wind farm is required by transmission system operators to ensure predictable stable behaviour during system faults. In particular, induction generators cannot support the system voltage during faults, unlike steam or hydro turbine-driven synchronous generators.

Induction generators aren't used in current turbines. Instead, most turbines use variable speed generators combined with partial- or full-scale power converter between the turbine generator and the collector system, which generally have more desirable properties for grid interconnection and have Low voltage ride through-capabilities. Modern concepts use either doubly fed machines with partial-scale converters or squirrel-cage induction generators or synchronous generators (both permanently and electrically excited) with full scale converters.

Transmission systems operators will supply a wind farm developer with a grid code to specify the requirements for interconnection to the transmission grid. This will include power factor, constancy of frequency and dynamic behaviour of the wind farm turbines during a system fault.

## Offshore Wind Power

The world's second full-scale floating wind turbine (and first to be installed without the use of heavy-lift vessels), WindFloat, operating at rated capacity (2 MW) approximately 5 km offshore of Póvoa de Varzim, Portugal

Offshore wind power refers to the construction of wind farms in large bodies of water to generate electric power. These installations can utilize the more frequent and powerful winds that are available in these locations and have less aesthetic impact on the landscape than land based projects. However, the construction and the maintenance costs are considerably higher.

Siemens and Vestas are the leading turbine suppliers for offshore wind power. Ørsted, Vattenfall and E.ON are the leading offshore operators. As of October 2010, 3.16 GW of offshore wind power capacity was operational, mainly in Northern Europe. Offshore wind power capacity is expected to reach a total of 75 GW worldwide by 2020, with significant contributions from China and the US. The UK's investments in offshore wind power have resulted in a rapid decrease of the usage of coal as an energy source between 2012 and 2017, as well as a drop in the usage of natural gas as an energy source in 2017.

In 2012, 1,662 turbines at 55 offshore wind farms in 10 European countries produced 18 TWh, enough to power almost five million households. As of September 2018 the Walney Extension in the United Kingdom is the largest offshore wind farm in the world at 659 MW.

## Collection and Transmission Network

In a wind farm, individual turbines are interconnected with a medium voltage (usually 34.5 kV) power collection system and communications network. At a substation, this medium-voltage electric current is increased in voltage with a transformer for connection to the high voltage electric power transmission system.

A transmission line is required to bring the generated power to (often remote) markets. For an off-shore station this may require a submarine cable. Construction of a new high-voltage line may be too costly for the wind resource alone, but wind sites may take advantage of lines installed for conventionally fueled generation.

One of the biggest current challenges to wind power grid integration in the United States is the necessity of developing new transmission lines to carry power from wind farms, usually in remote lowly populated states in the middle of the country due to availability of wind, to high load locations, usually on the coasts where population density is higher. The current transmission lines in remote locations were not designed for the transport of large amounts of energy. As transmission lines become longer the losses associated with power transmission increase, as modes of losses at lower lengths are exacerbated and new modes of losses are no longer negligible as the length is increased, making it harder to transport large loads over large distances. However, resistance from state and local governments makes it difficult to construct new transmission lines. Multi state power transmission projects are discouraged by states with cheap electric power rates for fear that exporting their cheap power will lead to increased rates. A 2005 energy law gave the Energy Department authority to approve transmission projects states refused to act on, but after an attempt to use this authority, the Senate declared the department was being overly aggressive in doing so. Another problem is that wind companies find out after the fact that the transmission capacity of a new farm is below the generation capacity, largely because federal utility rules to encourage renewable energy installation allow feeder lines to meet only minimum standards. These are important issues that need to be solved, as when the transmission capacity does not meet the generation capacity, wind farms are forced to produce below their full potential or stop running all together, in a process known as curtailment. While this leads to potential renewable generation left untapped, it prevents possible grid overload or risk to reliable service.

## Wind Power Capacity and Production

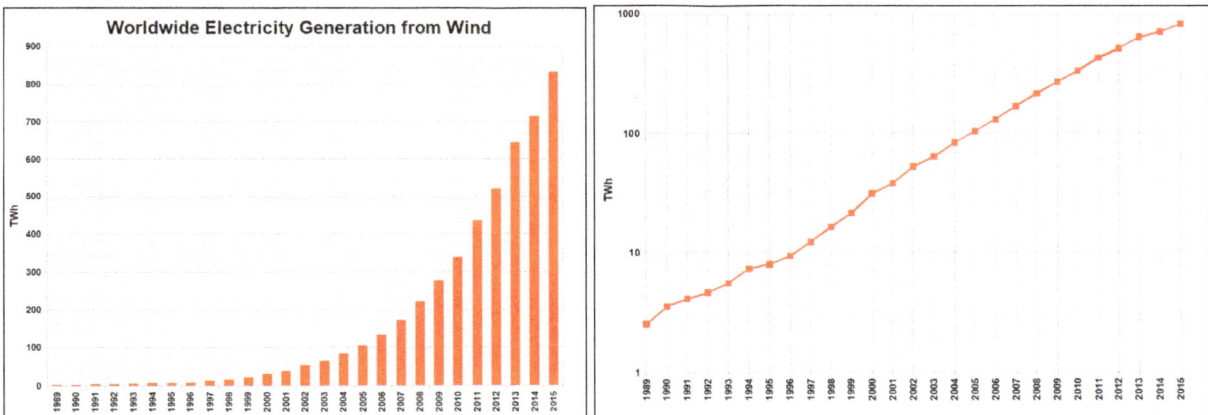

Worldwide wind generation.

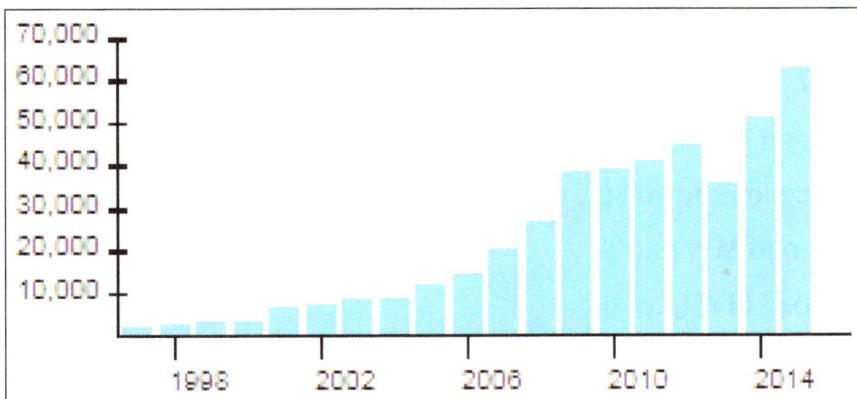

Global annual new installed wind capacity.

In 2015 there were over 200,000 wind turbines operating, with a total nameplate capacity of 432 GW worldwide. The European Union passed 100 GW nameplate capacity in September 2012, while the United States surpassed 75 GW in 2015 and China's grid connected capacity passed 145 GW in 2015. In 2015 wind power constituted 15.6% of all installed power generation capacity in the European Union and it generated around 11.4% of its power.

World wind generation capacity more than quadrupled between 2000 and 2006, doubling about every 3 years. The United States pioneered wind farms and led the world in installed capacity in the 1980s and into the 1990s. In 1997 installed capacity in Germany surpassed the United States and led until once again overtaken by the United States in 2008. China has been rapidly expanding its wind installations in the late 2000s and passed the United States in 2010 to become the world leader. As of 2011, 83 countries around the world were using wind power on a commercial basis.

The actual amount of electric power that wind is able to generate is calculated by multiplying the nameplate capacity by the capacity factor, which varies according to equipment and location. Estimates of the capacity factors for wind installations are in the range of 35% to 44%.

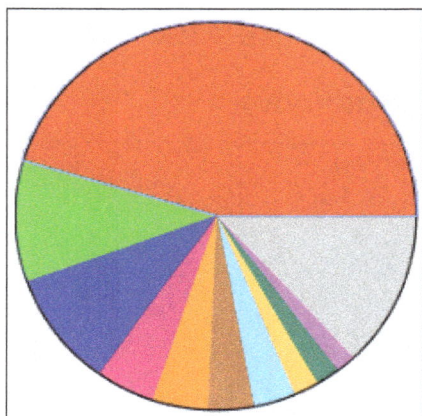

- ◼ China: 23,351 MW (45.4%).
- ◼ Germany: 5,279 MW (10.3%).
- ◼ United States: 4,854 MW (9.4%).
- ◼ Brazil: 2,472 MW (4.8%).
- ◼ India: 2,315 MW (4.5%).
- ◼ Canada: 1,871 MW (3.6%).
- ◼ United Kingdom: 1,736 MW (3.4%).
- ◼ Sweden: 1,050 MW (2.0%).
- ◼ France: 1,042 MW (2.0%).
- ◼ Turkey: 804 MW (1.6%).
- ◼ Rest of the world: 6,702 MW (13.0%).

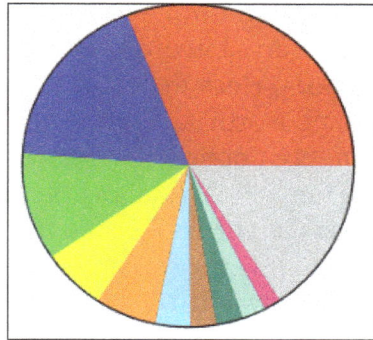

- ■ China: 114,763 MW (31.1%).

- ■ United States: 65,879 MW (17.8%).

- ■ Germany: 39,165 MW (10.6%).

- ■ Spain: 22,987 MW (6.2%).

- ■ India: 22,465 MW (6.1%).

- ■ United Kingdom: 12,440 MW (3.4%).

- ■ Canada: 9,694 MW (2.6%).

- ■ France: 9,285 MW (2.5%).

- ■ Italy: 8,663 MW (2.3%).

- ■ Brazil: 5,939 MW (1.6%).

- ■ Rest of the world: 58,275 MW (15.8%).

## Growth Trends

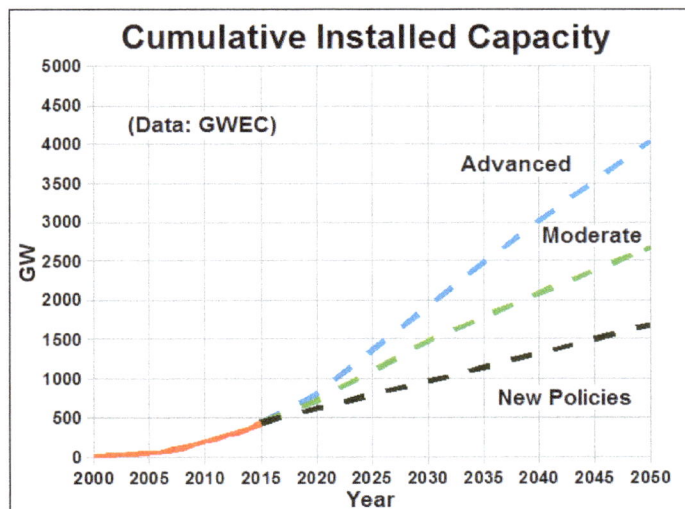

Worldwide installed wind power capacity forecast.

The wind power industry set new records in 2014 – more than 50 GW of new capacity was installed. Another record breaking year occurred in 2015, with 22% annual market growth resulting in the 60 GW mark being passed. In 2015, close to half of all new wind power was added outside of the traditional markets in Europe and North America. This was largely from new construction in China and India. Global Wind Energy Council (GWEC) figures show that 2015 recorded an increase of installed capacity of more than 63 GW, taking the total installed wind energy capacity to 432.9 GW, up from 74 GW in 2006. In terms of economic value, the wind energy sector has become one of the important players in the energy markets, with the total investments reaching US$329bn (€296.6bn), an increase of 4% over 2014.

Although the wind power industry was affected by the global financial crisis in 2009 and 2010, GWEC predicts that the installed capacity of wind power will be 792.1 GW by the end of 2020 and 4,042 GW by end of 2050. The increased commissioning of wind power is being accompanied by record low prices for forthcoming renewable electric power. In some cases, wind onshore is already the cheapest electric power generation option and costs are continuing to decline. The contracted prices for wind onshore for the next few years are now as low as 30 USD/MWh.

In the EU in 2015, 44% of all new generating capacity was wind power; while in the same period net fossil fuel power capacity decreased.

## Capacity Factor

Since wind speed is not constant, a wind farm's annual energy production is never as much as the sum of the generator nameplate ratings multiplied by the total hours in a year. The ratio of actual productivity in a year to this theoretical maximum is called the capacity factor. Typical capacity factors are 15–50%; values at the upper end of the range are achieved in favourable sites and are due to wind turbine design improvements.

Online data is available for some locations, and the capacity factor can be calculated from the yearly output. For example, the German nationwide average wind power capacity factor over all of 2012 was just under 17.5% (45,867 GW·h/yr / (29.9 GW × 24 × 366) = 0.1746), and the capacity factor for Scottish wind farms averaged 24% between 2008 and 2010.

Unlike fueled generating plants, the capacity factor is affected by several parameters, including the variability of the wind at the site and the size of the generator relative to the turbine's swept area. A small generator would be cheaper and achieve a higher capacity factor but would produce less electric power (and thus less profit) in high winds. Conversely, a large generator would cost more but generate little extra power and, depending on the type, may stall out at low wind speed. Thus an optimum capacity factor of around 40–50% would be aimed for.

A 2008 study released by the U.S. Department of Energy noted that the capacity factor of new wind installations was increasing as the technology improves, and projected further improvements for future capacity factors. In 2010, the department estimated the capacity factor of new wind turbines in 2010 to be 45%. The annual average capacity factor for wind generation in the US has varied between 29.8% and 34% during the period 2010–2015.

## Penetration

| Country | Penetration (Percentage of wind power generation over total electricity consumption) |
|---|---|
| Denmark (2015) | 42% |
| Portugal (2013) | 23% |
| Ireland (2015) | 23% |
| Spain (2014) | 20.2% |
| Germany (2017) | 18.7% |
| United Kingdom (2018) | 18% |
| United States (2016) | 6% |

Wind energy penetration is the fraction of energy produced by wind compared with the total generation. The wind power penetration in world electric power generation in 2015 was 3.5%.

There is no generally accepted maximum level of wind penetration. The limit for a particular grid will depend on the existing generating plants, pricing mechanisms, capacity for energy storage, demand management and other factors. An interconnected electric power grid will already include reserve generating and transmission capacity to allow for equipment failures. This reserve capacity can also serve to compensate for the varying power generation produced by wind stations. Studies have indicated that 20% of the total annual electrical energy consumption may be incorporated with minimal difficulty. These studies have been for locations with geographically dispersed wind farms, some degree of dispatchable energy or hydropower with storage capacity, demand management, and interconnected to a large grid area enabling the export of electric power when needed. Beyond the 20% level, there are few technical limits, but the economic implications become more significant. Electrical utilities continue to study the effects of large scale penetration of wind generation on system stability and economics.

A wind energy penetration figure can be specified for different duration of time, but is often quoted annually. To obtain 100% from wind annually requires substantial long term storage or substantial interconnection to other systems which may already have substantial storage. On a monthly, weekly, daily, or hourly basis—or less—wind might supply as much as or more than 100% of current use, with the rest stored or exported. Seasonal industry might then take advantage of high wind and low usage times such as at night when wind output can exceed normal demand. Such industry might include production of silicon, aluminum, steel, or of natural gas, and hydrogen, and using future long term storage to facilitate 100% energy from variable renewable energy. Homes can also be programmed to accept extra electric power on demand, for example by remotely turning up water heater thermostats.

In Australia, the state of South Australia generates around half of the nation's wind power capacity. By the end of 2011 wind power in South Australia, championed by Premier (and Climate Change Minister) Mike Rann, reached 26% of the State's electric power generation, edging out coal for the first time. At this stage South Australia, with only 7.2% of Australia's population, had 54% of Australia's installed capacity.

## Variability

Wind turbines are typically installed in favorable windy locations.

Electric power generated from wind power can be highly variable at several different timescales: hourly, daily, or seasonally. Annual variation also exists, but is not as significant. Because instantaneous electrical generation and consumption must remain in balance to maintain grid stability, this variability can present substantial challenges to incorporating large amounts of wind power into a grid system. Intermittency and the non-dispatchable nature of wind energy production can raise costs for regulation, incremental operating reserve, and (at high penetration levels) could require an increase in the already existing energy demand management, load shedding, storage solutions or system interconnection with HVDC cables.

Fluctuations in load and allowance for failure of large fossil-fuel generating units requires operating reserve capacity, which can be increased to compensate for variability of wind generation.

Wind power is variable, and during low wind periods it must be replaced by other power sources. Transmission networks presently cope with outages of other generation plants and daily changes in electrical demand, but the variability of intermittent power sources such as wind power, is more frequent than those of conventional power generation plants which, when scheduled to be operating, may be able to deliver their nameplate capacity around 95% of the time.

Presently, grid systems with large wind penetration require a small increase in the frequency of usage of natural gas spinning reserve power plants to prevent a loss of electric power in the event that there is no wind. At low wind power penetration, this is less of an issue.

GE has installed a prototype wind turbine with onboard battery similar to that of an electric car, equivalent of 1 minute of production. Despite the small capacity, it is enough to guarantee that power output complies with forecast for 15 minutes, as the battery is used to eliminate the difference rather than provide full output. In certain cases the increased predictability can be used to take wind power penetration from 20 to 30 or 40 per cent. The battery cost can be retrieved by selling burst power on demand and reducing backup needs from gas plants.

In the UK there were 124 separate occasions from 2008 to 2010 when the nation's wind output fell to less than 2% of installed capacity. A report on Denmark's wind power noted that their wind power network provided less than 1% of average demand on 54 days during the year 2002. Wind power advocates argue that these periods of low wind can be dealt with by simply restarting existing power stations that have been held in readiness, or interlinking with HVDC. Electrical grids with slow-responding thermal power plants and without ties to networks with hydroelectric generation may have to limit the use of wind power. According to a 2007 Stanford University study published in the Journal of Applied Meteorology and Climatology, interconnecting ten or more wind farms can allow an average of 33% of the total energy produced (i.e. about 8% of total nameplate capacity) to be used as reliable, baseload electric power which can be relied on to handle peak loads, as long as minimum criteria are met for wind speed and turbine height.

Conversely, on particularly windy days, even with penetration levels of 16%, wind power generation can surpass all other electric power sources in a country. In Spain, in the early hours of 16 April 2012 wind power production reached the highest percentage of electric power production till then, at 60.46% of the total demand. In Denmark, which had power market penetration of 30% in 2013, over 90 hours, wind power generated 100% of the country's power, peaking at 122% of the country's demand at 2 am on 28 October.

Table: Increase in system operation costs, euros per MWH, for 10% and 20% wind share.

| Country | 10% | 20% |
|---------|-----|-----|
| Germany | 2.5 | 3.2 |
| Denmark | 0.4 | 0.8 |
| Finland | 0.3 | 1.5 |
| Norway  | 0.1 | 0.3 |
| Sweden  | 0.3 | 0.7 |

A 2006 International Energy Agency forum presented costs for managing intermittency as a function of wind-energy's share of total capacity for several countries, as shown in the table on the right. Three reports on the wind variability in the UK issued in 2009, generally agree that variability of wind needs to be taken into account by adding 20% to the operating reserve, but it does not make the grid unmanageable. The additional costs, which are modest, can be quantified.

The combination of diversifying variable renewables by type and location, forecasting their variation, and integrating them with dispatchable renewables, flexible fueled generators, and demand response can create a power system that has the potential to meet power supply needs reliably. Integrating ever-higher levels of renewables is being successfully demonstrated in the real world:

In 2009, eight American and three European authorities, writing in the leading electrical engineers' professional journal, didn't find "a credible and firm technical limit to the amount of wind energy that can be accommodated by electric power grids". In fact, not one of more than 200 international studies, nor official studies for the eastern and western U.S. regions, nor the International Energy Agency, has found major costs or technical barriers to reliably integrating up to 30% variable renewable supplies into the grid, and in some studies much more.

Seasonal cycle of capacity factors for wind and photovoltaics in Europe under idealized assumptions. Solar power tends to be complementary to wind. On daily to weekly timescales, high pressure areas tend to bring clear skies and low surface winds, whereas low pressure areas tend to be windier and cloudier. On seasonal timescales, solar energy peaks in summer, whereas in many areas wind energy is lower in summer and higher in winter. Thus the seasonal variation of wind and solar power tend to cancel each other somewhat. In 2007 the Institute for Solar Energy Supply Technology of the University of Kassel pilot-tested a combined power plant linking solar, wind, biogas and hydrostorage to provide load-following power around the clock and throughout the year, entirely from renewable sources.

## Predictability

Wind power forecasting methods are used, but predictability of any particular wind farm is low for short-term operation. For any particular generator there is an 80% chance that wind output will change less than 10% in an hour and a 40% chance that it will change 10% or more in 5 hours.

However, studies by Graham Sinden suggest that, in practice, the variations in thousands of wind turbines, spread out over several different sites and wind regimes, are smoothed. As the distance between sites increases, the correlation between wind speeds measured at those sites, decreases.

Thus, while the output from a single turbine can vary greatly and rapidly as local wind speeds vary, as more turbines are connected over larger and larger areas the average power output becomes less variable and more predictable.

Wind power hardly ever suffers major technical failures, since failures of individual wind turbines have hardly any effect on overall power, so that the distributed wind power is reliable and predictable, whereas conventional generators, while far less variable, can suffer major unpredictable outages.

## Energy Storage

The Sir Adam Beck Generating Complex at Niagara Falls, Canada, includes a large pumped-storage hydroelectricity reservoir. During hours of low electrical demand excess electrical grid power is used to pump water up into the reservoir, which then provides an extra 174 MW of electric power during periods of peak demand.

Typically, conventional hydroelectricity complements wind power very well. When the wind is blowing strongly, nearby hydroelectric stations can temporarily hold back their water. When the wind drops they can, provided they have the generation capacity, rapidly increase production to compensate. This gives a very even overall power supply and virtually no loss of energy and uses no more water.

Alternatively, where a suitable head of water is not available, pumped-storage hydroelectricity or other forms of grid energy storage such as compressed air energy storage and thermal energy storage can store energy developed by high-wind periods and release it when needed. The type of storage needed depends on the wind penetration level – low penetration requires daily storage, and high penetration requires both short and long term storage – as long as a month or more. Stored energy increases the economic value of wind energy since it can be shifted to displace higher cost generation during peak demand periods. The potential revenue from this arbitrage can offset the cost and losses of storage. For example, in the UK, the 1.7 GW Dinorwig pumped-storage plant evens out electrical demand peaks, and allows base-load suppliers to run their plants more efficiently. Although pumped-storage power systems are only about 75% efficient, and have high installation costs, their low running costs and ability to reduce the required electrical base-load can save both fuel and total electrical generation costs.

In particular geographic regions, peak wind speeds may not coincide with peak demand for electrical power. In the U.S. states of California and Texas, for example, hot days in summer may have low wind speed and high electrical demand due to the use of air conditioning. Some utilities subsidize the purchase of geothermal heat pumps by their customers, to reduce electric power demand during the summer months by making air conditioning up to 70% more efficient; widespread adoption of this technology would better match electric power demand to wind availability in areas with hot summers and low summer winds. A possible future option may be to interconnect widely dispersed geographic areas with an HVDC "super grid". In the U.S. it is estimated that to upgrade the transmission system to take in planned or potential renewables would cost at least USD 60 bn, while the society value of added windpower would be more than that cost.

Germany has an installed capacity of wind and solar that can exceed daily demand, and has been exporting peak power to neighboring countries, with exports which amounted to some 14.7 billion kWh in 2012. A more practical solution is the installation of thirty days storage capacity able to supply 80% of demand, which will become necessary when most of Europe's energy is obtained from wind power and solar power. Just as the EU requires member countries to maintain 90 days strategic reserves of oil it can be expected that countries will provide electric power storage, instead of expecting to use their neighbors for net metering.

## Capacity Credit, Fuel Savings and Energy Payback

The capacity credit of wind is estimated by determining the capacity of conventional plants displaced by wind power, whilst maintaining the same degree of system security. According to the American Wind Energy Association, production of wind power in the United States in 2015 avoided consumption of $2.8 \times 10^8$ cubic metres ($73 \times 10^9$ US gallons) of water and reduced $CO_2$ emissions by 132 million metric tons, while providing USD 7.3 bn in public health savings.

The energy needed to build a wind farm divided into the total output over its life, Energy Return on Energy Invested, of wind power varies but averages about 20–25. Thus, the energy payback time is typically around a year.

## Economics

According to BusinessGreen, wind turbines reached grid parity (the point at which the cost of wind power matches traditional sources) in some areas of Europe in the mid-2000s, and in the US around the same time. Falling prices continue to drive the levelized cost down and it has been suggested that it has reached general grid parity in Europe in 2010, and will reach the same point in the US around 2016 due to an expected reduction in capital costs of about 12%. According to PolitiFact, it is difficult to predict whether wind power would remain viable in the United States without subsidies.

## Electric Power Cost and Trends

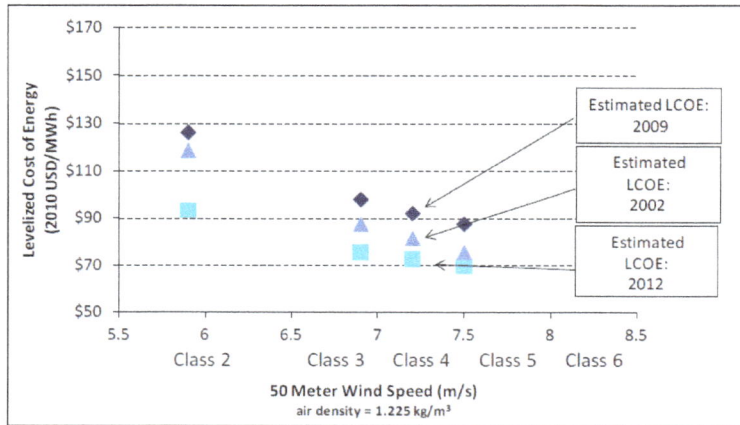

Estimated cost per MWh for wind power.

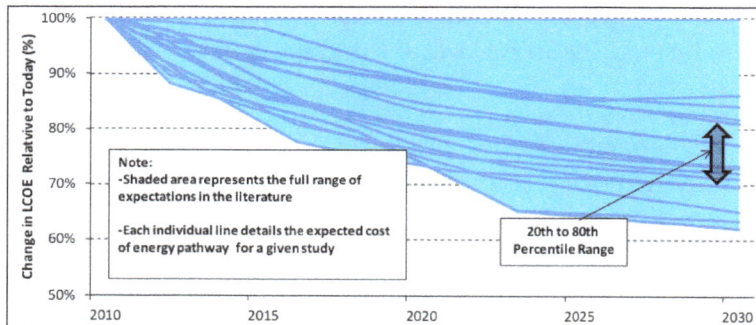

A turbine blade convoy passing through Edenfield in the U.K. Even longer two-piece blades are now manufactured, and then assembled on-site to reduce difficulties in transportation.

Wind power is capital intensive, but has no fuel costs. The price of wind power is therefore much more stable than the volatile prices of fossil fuel sources. The marginal cost of wind energy once a station is constructed is usually less than 1-cent per kW·h.

However, the estimated average cost per unit of electric power must incorporate the cost of construction of the turbine and transmission facilities, borrowed funds, return to investors (including cost of risk), estimated annual production, and other components, averaged over the projected useful life of the equipment, which may be in excess of twenty years. Energy cost estimates are highly dependent on these assumptions so published cost figures can differ substantially. In 2004, wind energy cost a fifth of what it did in the 1980s, and some expected that downward trend to continue as larger multi-mega-watt turbines were mass-produced. In 2012 capital costs for wind turbines were substantially lower than 2008–2010 but still above 2002 levels. A 2011 report from the American Wind Energy Association stated, "Wind's costs have dropped over the past two years, in the range of 5 to 6 cents per kilowatt-hour recently.... about 2 cents cheaper than coal-fired electric power, and more projects were financed through debt arrangements than tax equity structures last year.... winning more mainstream acceptance from Wall Street's banks. Equipment makers can also deliver products in the same year that they are ordered instead of waiting up to three years as was the case in previous cycles. 5,600 MW of new installed capacity is under construction in the United States, more than double the number at this point in 2010. Thirty-five percent of all new power generation built in the United States since 2005 has come from wind, more than new gas and coal plants combined, as power providers are increasingly enticed to wind as a convenient hedge against unpredictable commodity price moves".

A British Wind Energy Association report gives an average generation cost of onshore wind power of around 3.2 pence (between US 5 and 6 cents) per kW·h. Cost per unit of energy produced was estimated in 2006 to be 5 to 6 percent above the cost of new generating capacity in the US for coal and natural gas: wind cost was estimated at $55.80 per MW·h, coal at $53.10/MW·h and natural gas at $52.50. Similar comparative results with natural gas were obtained in a governmental study in the UK in 2011. In 2011 power from wind turbines could be already cheaper than fossil or nuclear plants; it is also expected that wind power will be the cheapest form of energy generation in the future. The presence of wind energy, even when subsidised, can reduce costs for consumers (€5 billion/yr in Germany) by reducing the marginal price, by minimising the use of expensive peaking power plants.

A 2012 EU study shows base cost of onshore wind power similar to coal, when subsidies and externalities are disregarded. Wind power has some of the lowest external costs.

In February 2013 Bloomberg New Energy Finance (BNEF) reported that the cost of generating electric power from new wind farms is cheaper than new coal or new baseload gas plants. When including the current Australian federal government carbon pricing scheme their modeling gives costs (in Australian dollars) of $80/MWh for new wind farms, $143/MWh for new coal plants and $116/MWh for new baseload gas plants. The modeling also shows that "even without a carbon price (the most efficient way to reduce economy-wide emissions) wind energy is 14% cheaper than new coal and 18% cheaper than new gas." Part of the higher costs for new coal plants is due to high financial lending costs because of "the reputational damage of emissions-intensive investments". The expense of gas fired plants is partly due to "export market" effects on local prices. Costs of production from coal fired plants built in "the 1970s and 1980s" are cheaper than renewable energy sources because of depreciation. In 2015 BNEF calculated LCOE prices per MWh energy in new powerplants (excluding carbon costs) : $85 for onshore wind ($175 for offshore), $66–75 for coal in the Americas ($82–105 in Europe), gas $80–100. A 2014 study showed unsubsidized LCOE costs between $37–81, depending on region. A 2014 US DOE report showed that in some cases power purchase agreement prices for wind power had dropped to record lows of $23.5/MWh.

The cost has reduced as wind turbine technology has improved. There are now longer and lighter wind turbine blades, improvements in turbine performance and increased power generation efficiency. Also, wind project capital and maintenance costs have continued to decline. For example, the wind industry in the US in early 2014 were able to produce more power at lower cost by using taller wind turbines with longer blades, capturing the faster winds at higher elevations. This has opened up new opportunities and in Indiana, Michigan, and Ohio, the price of power from wind turbines built 90–120 metres (300–400 ft) above the ground can since 2014 compete with conventional fossil fuels like coal. Prices have fallen to about 4 cents per kilowatt-hour in some cases and utilities have been increasing the amount of wind energy in their portfolio, saying it is their cheapest option.

A number of initiatives are working to reduce costs of electric power from offshore wind. One example is the Carbon Trust Offshore Wind Accelerator, a joint industry project, involving nine offshore wind developers, which aims to reduce the cost of offshore wind by 10% by 2015. It has been suggested that innovation at scale could deliver 25% cost reduction in offshore wind by 2020. Henrik Stiesdal, former Chief Technical Officer at Siemens Wind Power, has stated that by 2025 energy from offshore wind will be one of the cheapest, scalable solutions in the UK, compared to other renewables and fossil fuel energy sources, if the true cost to society is factored into the cost of energy equation. He calculates the cost at that time to be 43 EUR/MWh for onshore, and 72 EUR/MWh for offshore wind.

In August 2017, the Department of Energy's National Renewable Energy Laboratory (NREL) published a new report on a 50% reduction in wind power cost by 2030. The NREL is expected to achieve advances in wind turbine design, materials and controls to unlock performance improvements and reduce costs. According to international surveyors, this study shows that cost cutting is projected to fluctuate between 24% and 30% by 2030. In more aggressive cases, experts estimate cost reduction Up to 40 percent if the research and development and technology programs result in additional efficiency.

In 2018 a Lazard study found that "The low end levelized cost of onshore wind-generated energy is $29/MWh, compared to an average illustrative marginal cost of $36/MWh for coal", and noted that the average cost had fallen by 7% in a year.

## Incentives and Community Benefits

U.S. landowners typically receive $3,000–$5,000 annual rental income per wind turbine, while farmers continue to grow crops or graze cattle up to the foot of the turbines. Shown: the Brazos Wind Farm, Texas.

Some of the 6,000 turbines in California's Altamont Pass Wind Farm aided by tax incentives during the 1980s.

The wind industry in the United States generates tens of thousands of jobs and billions of dollars of economic activity. Wind projects provide local taxes, or payments in lieu of taxes and strengthen the economy of rural communities by providing income to farmers with wind turbines on their land. Wind energy in many jurisdictions receives financial or other support to encourage its development. Wind energy benefits from subsidies in many jurisdictions, either to increase its attractiveness, or to compensate for subsidies received by other forms of production which have significant negative externalities.

In the US, wind power receives a production tax credit (PTC) of 1.5¢/kWh in 1993 dollars for each kW·h produced, for the first ten years; at 2.2 cents per kW·h in 2012, the credit was renewed on 2 January 2012, to include construction begun in 2013. A 30% tax credit can be applied instead of receiving the PTC. Another tax benefit is accelerated depreciation. Many American states also provide incentives, such as exemption from property tax, mandated purchases, and additional markets for "green credits". The Energy Improvement and Extension Act of 2008 contains extensions of credits for wind, including microturbines. Countries such as Canada and Germany also provide incentives for wind turbine construction, such as tax credits or minimum purchase prices for wind generation, with assured grid access (sometimes referred to as feed-in tariffs). These feed-in tariffs are typically set well above average electric power prices. In December 2013 U.S. Senator Lamar Alexander and other Republican senators argued that the "wind energy production tax credit should be allowed to expire at the end of 2013" and it expired 1 January 2014 for new installations.

Secondary market forces also provide incentives for businesses to use wind-generated power, even if there is a premium price for the electricity. For example, socially responsible manufacturers pay utility companies a premium that goes to subsidize and build new wind power infrastructure. Companies use wind-generated power, and in return they can claim that they are undertaking strong "green" efforts. In the US the organization Green-e monitors business compliance with these renewable energy credits. Turbine prices have fallen significantly in recent years due to tougher competitive conditions such as the increased use of energy auctions, and the elimination of subsidies in many markets. For example, Vestas, a wind turbine manufacturer, whose largest onshore turbine can pump out 4.2 megawatts of power, enough to provide electricity to roughly 5,000 homes, has seen prices for its turbines fall from €950,000 per megawatt in late 2016, to around €800,000 per megawatt in the third quarter of 2017.

## Small-scale Wind Power

A small Quietrevolution QR5 Gorlov type vertical axis wind turbine on the roof of Colston Hall in Bristol, England. Measuring 3 m in diameter and 5 m high, it has a nameplate rating of 6.5 kW.

Small-scale wind power is the name given to wind generation systems with the capacity to produce up to 50 kW of electrical power. Isolated communities, that may otherwise rely on diesel generators, may use wind turbines as an alternative. Individuals may purchase these systems to reduce or eliminate their dependence on grid electric power for economic reasons, or to reduce their carbon footprint. Wind turbines have been used for household electric power generation in conjunction with battery storage over many decades in remote areas.

Recent examples of small-scale wind power projects in an urban setting can be found in New York City, where, since 2009, a number of building projects have capped their roofs with Gorlov-type helical wind turbines. Although the energy they generate is small compared to the buildings' overall consumption, they help to reinforce the building's 'green' credentials in ways that showing people your high-tech boiler cannot, with some of the projects also receiving the direct support of the New York State Energy Research and Development Authority.

Grid-connected domestic wind turbines may use grid energy storage, thus replacing purchased electric power with locally produced power when available. The surplus power produced by domestic microgenerators can, in some jurisdictions, be fed into the network and sold to the utility company, producing a retail credit for the microgenerators' owners to offset their energy costs.

Off-grid system users can either adapt to intermittent power or use batteries, photovoltaic or diesel systems to supplement the wind turbine. Equipment such as parking meters, traffic warning signs, street lighting, or wireless Internet gateways may be powered by a small wind turbine, possibly combined with a photovoltaic system, that charges a small battery replacing the need for a connection to the power grid.

A Carbon Trust study into the potential of small-scale wind energy in the UK, published in 2010, found that small wind turbines could provide up to 1.5 terawatt hours (TW·h) per year of electric power (0.4% of total UK electric power consumption), saving 0.6 million tonnes of carbon dioxide

(Mt CO₂) emission savings. This is based on the assumption that 10% of households would install turbines at costs competitive with grid electric power, around 12 pence (US 19 cents) a kW·h. A report prepared for the UK's government-sponsored Energy Saving Trust in 2006, found that home power generators of various kinds could provide 30 to 40% of the country's electric power needs by 2050.

Distributed generation from renewable resources is increasing as a consequence of the increased awareness of climate change. The electronic interfaces required to connect renewable generation units with the utility system can include additional functions, such as the active filtering to enhance the power quality.

## Environmental Effects

Livestock grazing near a wind turbine.

The environmental impact of wind power when compared to the environmental impacts of fossil fuels, is relatively minor. According to the IPCC, in assessments of the life-cycle global warming potential of energy sources, wind turbines have a median value of between 12 and 11 (gCO2eq/ kWh) depending on whether off- or onshore turbines are being assessed. Compared with other low carbon power sources, wind turbines have some of the lowest global warming potential per unit of electrical energy generated.

Onshore wind farms can have a significant visual impact and impact on the landscape. Their network of turbines, access roads, transmission lines and substations can result in "energy sprawl". Typically they need to cover more land and be more spread out than other power stations. To power many major cities by wind alone would require building wind farms bigger than the cities themselves. Typically they also need to be built in wild and rural areas, which can lead to "industrialization of the countryside" and habitat loss. A report by the Mountaineering Council of Scotland concluded that wind farms have a negative impact on tourism in areas known for natural landscapes and panoramic views. However, land between the turbines and roads can still be used for agriculture.

Habitat loss and habitat fragmentation are the greatest impact of wind farms on wildlife. There are also reports of higher bird and bat mortality at wind turbines as there are around other artificial structures. The scale of the ecological impact may or may not be significant, depending on specific circumstances. Prevention and mitigation of wildlife fatalities, and protection of peat bogs, affect the siting and operation of wind turbines.

Wind turbines generate some noise. At a residential distance of 300 metres (980 ft) this may be around 45 dB, which is slightly louder than a refrigerator. At 1.5 km (1 mi) distance they become inaudible. There are anecdotal reports of negative health effects from noise on people who live very close to wind turbines. Peer-reviewed research has generally not supported these claims.

The United States Air Force and Navy have expressed concern that siting large wind turbines near bases "will negatively impact radar to the point that air traffic controllers will lose the location of aircraft".

Before 2019, many wind turbine blades had been made of fiberglass with designs that only provided a service lifetime of 10 to 20 years. Given the available technology, As of February 2018 there was no market for recycling these old blades. One common disposal option was to truck them to landfills. Because they were designed to be hollow took up enormous volume compared to their mass. Landfill operators have started requiring operators to crush the blades before they can be landfilled.

## Politics

## Central Government

Part of the Seto Hill Windfarm in Japan.

Nuclear power and fossil fuels are subsidized by many governments, and wind power and other forms of renewable energy are also often subsidized. For example, a 2009 study by the Environmental Law Institute assessed the size and structure of U.S. energy subsidies over the 2002–2008 period. The study estimated that subsidies to fossil-fuel based sources amounted to approximately $72 billion over this period and subsidies to renewable fuel sources totalled $29 billion. In the United States, the federal government has paid US$74 billion for energy subsidies to support R&D for nuclear power ($50 billion) and fossil fuels ($24 billion) from 1973 to 2003. During this same time frame, renewable energy technologies and energy efficiency received a total of US$26 billion. It has been suggested that a subsidy shift would help to level the playing field and support growing energy sectors, namely solar power, wind power, and biofuels. History shows that no energy sector was developed without subsidies.

According to the International Energy Agency (IEA), energy subsidies artificially lower the price of energy paid by consumers, raise the price received by producers or lower the cost of production. "Fossil fuels subsidies costs generally outweigh the benefits. Subsidies to renewables and low-carbon energy technologies can bring long-term economic and environmental benefits". In November 2011, an IEA report entitled Deploying Renewables 2011 said "subsidies in green energy technologies that were not yet competitive are justified in order to give an incentive to investing into technologies with clear environmental and energy security benefits". The IEA's report disagreed with claims that renewable energy technologies are only viable through costly subsidies and not able to produce energy reliably to meet demand.

However, IEA's views are not universally accepted. Between 2010 and 2016, subsidies for wind were between 1.3¢ and 5.7¢ per kWh. Subsidies for coal, natural gas and nuclear are all between 0.05¢ and 0.2¢ per kWh over all years. On a per-kWh basis, wind is subsidized 50 times as much as traditional sources.

In the United States, the wind power industry has recently increased its lobbying efforts considerably, spending about $5 million in 2009 after years of relative obscurity in Washington. By comparison, the U.S. nuclear industry alone spent over $650 million on its lobbying efforts and campaign contributions during a ten-year period ending in 2008.

Following the 2011 Japanese nuclear accidents, Germany's federal government is working on a new plan for increasing energy efficiency and renewable energy commercialization, with a particular focus on offshore wind farms. Under the plan, large wind turbines will be erected far away from the coastlines, where the wind blows more consistently than it does on land, and where the enormous turbines won't bother the inhabitants. The plan aims to decrease Germany's dependence on energy derived from coal and nuclear power plants.

## Public Opinion

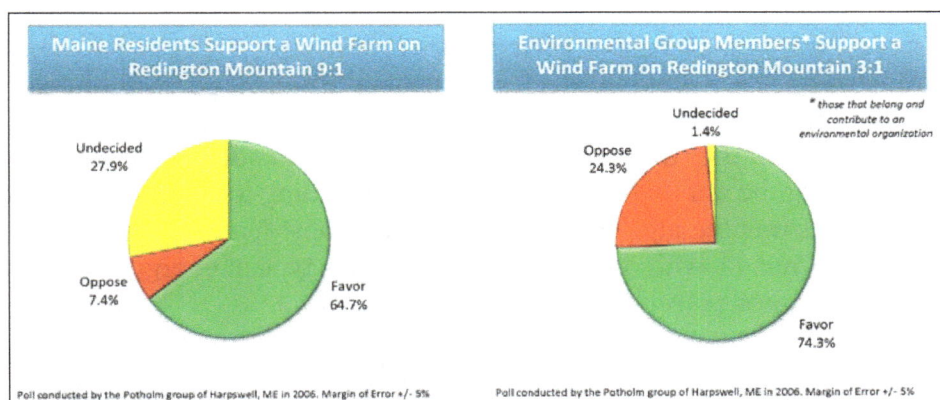

Environmental group members are both more in favor of wind power (74%) as well as more opposed (24%). Few are undecided.

Surveys of public attitudes across Europe and in many other countries show strong public support for wind power. About 80% of EU citizens support wind power. In Germany, where wind power has gained very high social acceptance, hundreds of thousands of people have invested in citizens' wind farms across the country and thousands of small and medium-sized enterprises are running successful businesses in a new sector that in 2008 employed 90,000 people and generated 8% of Germany's electric power.

Bakker discovered in their study that when residents did not want the turbines located by them their annoyance was significantly higher than those "that benefited economically from wind turbines the proportion of people who were rather or very annoyed was significantly lower".

Although wind power is a popular form of energy generation, the construction of wind farms is not universally welcomed, often for aesthetic reasons.

In Spain, with some exceptions, there has been little opposition to the installation of inland wind parks. However, the projects to build offshore parks have been more controversial. In particular, the proposal of building the biggest offshore wind power production facility in the world in southwestern Spain in the coast of Cádiz, on the spot of the 1805 Battle of Trafalgar has been met with strong opposition who fear for tourism and fisheries in the area, and because the area is a war grave.

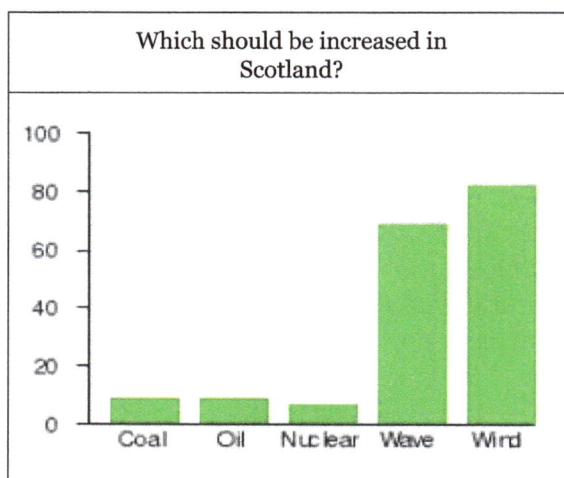

In a survey conducted by Angus Reid Strategies in October 2007, 89 per cent of respondents said that using renewable energy sources like wind or solar power was positive for Canada, because these sources were better for the environment. Only 4 per cent considered using renewable sources as negative since they can be unreliable and expensive. According to a Saint Consulting survey in April 2007, wind power was the alternative energy source most likely to gain public support for future development in Canada, with only 16% opposed to this type of energy. By contrast, 3 out of 4 Canadians opposed nuclear power developments.

A 2003 survey of residents living around Scotland's 10 existing wind farms found high levels of community acceptance and strong support for wind power, with much support from those who lived closest to the wind farms. The results of this survey support those of an earlier Scottish Executive survey 'Public attitudes to the Environment in Scotland 2002', which found that the Scottish public would prefer the majority of their electric power to come from renewables, and which rated wind power as the cleanest source of renewable energy. A survey conducted in 2005 showed that 74% of people in Scotland agree that wind farms are necessary to meet current and future energy needs. When people were asked the same question in a Scottish renewables study conducted in 2010, 78% agreed. The increase is significant as there were twice as many wind farms in 2010 as there were in 2005. The 2010 survey also showed that 52% disagreed with the statement that wind farms are "ugly and a blot on the landscape". 59% agreed that wind farms were necessary and that how they looked was unimportant. Regarding tourism, query responders consider power pylons, cell phone towers, quarries and plantations more negatively than wind farms. Scotland is planning to obtain 100% of electric power from renewable sources by 2020.

In other cases there is direct community ownership of wind farm projects. The hundreds of thousands of people who have become involved in Germany's small and medium-sized wind farms demonstrate such support there.

A 2010 Harris Poll reflects the strong support for wind power in Germany, other European countries, and the United States.

Opinion on increase in number of wind farms, 2010 Harris Poll.

|  | U.S. | Great Britain | France | Italy | Spain | Germany |
|---|---|---|---|---|---|---|
|  | % | % | % | % | % | % |
| Strongly oppose | 3 | 6 | 6 | 2 | 2 | 4 |
| Oppose more than favour | 9 | 12 | 16 | 11 | 9 | 14 |
| Favour more than oppose | 37 | 44 | 44 | 38 | 37 | 42 |
| Strongly favour | 50 | 38 | 33 | 49 | 53 | 40 |

In China, Shen discover that Chinese city-dwellers may be somewhat resistant to building wind turbines in urban areas, with a surprisingly high proportion of people citing an unfounded fear of radiation as driving their concerns. The central Chinese government rather than scientists is better suited to address this concern. In addition, the study finds that like their counterparts in OECD countries, urban Chinese respondents are sensitive to direct costs and to wildlife externalities. Distributing relevant information about turbines to the public may alleviate resistance.

## Community

Wind turbines such as these have been opposed for a number of reasons, including aesthetics, by some sectors of the population.

Many wind power companies work with local communities to reduce environmental and other concerns associated with particular wind farms. In other cases there is direct community ownership of wind farm projects. Appropriate government consultation, planning and approval procedures also help to minimize environmental risks. Some may still object to wind farms but, according to The Australia Institute, their concerns should be weighed against the need to address the threats posed by climate change and the opinions of the broader community.

In America, wind projects are reported to boost local tax bases, helping to pay for schools, roads and hospitals. Wind projects also revitalize the economy of rural communities by providing steady income to farmers and other landowners.

In the UK, both the National Trust and the Campaign to Protect Rural England have expressed concerns about the effects on the rural landscape caused by inappropriately sited wind turbines and wind farms.

A panoramic view.

Some wind farms have become tourist attractions. The Whitelee Wind Farm Visitor Centre has an exhibition room, a learning hub, a café with a viewing deck and also a shop. It is run by the Glasgow Science Centre.

In Denmark, a loss-of-value scheme gives people the right to claim compensation for loss of value of their property if it is caused by proximity to a wind turbine. The loss must be at least 1% of the property's value.

Despite this general support for the concept of wind power in the public at large, local opposition often exists and has delayed or aborted a number of projects. For example, there are concerns that some installations can negatively affect TV and radio reception and Doppler weather radar, as well as produce excessive sound and vibration levels leading to a decrease in property values. Potential broadcast-reception solutions include predictive interference modeling as a component of site selection. A study of 50,000 home sales near wind turbines found no statistical evidence that prices were affected.

While aesthetic issues are subjective and some find wind farms pleasant and optimistic, or symbols of energy independence and local prosperity, protest groups are often formed to attempt to block new wind power sites for various reasons.

This type of opposition is often described as NIMBYism, but research carried out in 2009 found that there is little evidence to support the belief that residents only object to renewable power facilities such as wind turbines as a result of a "Not in my Back Yard" attitude.

## Wind Farms

A wind farm or wind park, also called a wind power station or wind power plant, is a group of wind turbines in the same location used to produce electricity. A large wind farm may consist of several hundred individual wind turbines and cover an extended area of hundreds of square miles, but the land between the turbines may be used for agricultural or other purposes. A wind farm can also be located offshore.

Many of the largest operational onshore wind farms are located in China, India, and the United States. For example, the largest wind farm in the world, Gansu Wind Farm in China had a capacity of over 6,000 MW by 2012, with a goal of 20,000 MW by 2020. As of September 2018, the 659 MW Walney

Wind Farm in the UK is the largest offshore wind farm in the world. Individual wind turbine designs continue to increase in power, resulting in fewer turbines being needed for the same total output.

The San Gorgonio Pass wind farm.

Wind farms tend to have much less impact on the environment than many other power stations. Onshore wind farms are also criticized for their visual impact and impact on the landscape, as typically they need to take up more land than other power stations and need to be built in wild and rural areas, which can lead to "industrialization of the countryside", habitat loss, and a drop in tourism. Critics have linked wind farms to adverse health effects. Wind farms have also been criticized for interfering with radar, radio and television reception.

## Design and Location

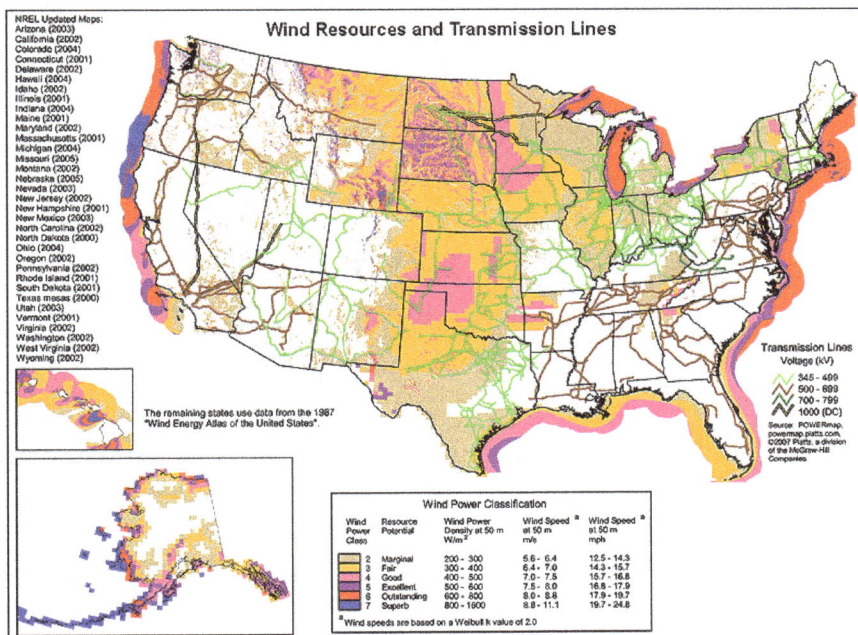

Map of available wind power over the United States. Color codes indicate wind power density class.

The location is critical to the success of a wind farm. Conditions contributing to a successful wind farm location include: wind conditions, access to electric transmission, physical access, and local electricity prices.

The faster the average wind speed, the more electricity the wind turbine will generate, so faster winds are generally economically better for wind farm developments. The balancing factor is that strong gusts and high turbulence require stronger more expensive turbines, otherwise they risk damage. The average power in the wind is not proportional to the average wind speed, however. For this reason, the ideal wind conditions would be strong but consistent winds with low turbulence coming from a single direction.

Mountain passes are ideal locations for wind farms under these conditions. Mountain passes channel wind blocked by mountains through a tunnel like pass towards areas of lower pressure and flatter land. Passes used for wind farms like the San Gorgonio Pass and Altamont Pass are known for their abundant wind resource capacity and capability for large-scale wind farms. These types of passes were the first places in the 1980s to have heavily invested large-scale wind farms after approval for wind energy development by the U.S. Bureau of Land Management. From these wind farms, developers learned a lot about turbulence and crowding effects of large-scale wind projects previously unresearched in the U.S. due to the lack of operational wind farms large enough to conduct these types of studies on.

Usually sites are screened on the basis of a wind atlas, and validated with on-site wind measurements via long term or permanent meteorological-tower data using anemometers and wind vanes. Meteorological wind data alone is usually not sufficient for accurate siting of a large wind power project. Collection of site specific data for wind speed and direction is crucial to determining site potential in order to finance the project. Local winds are often monitored for a year or more, detailed wind maps are constructed, along with rigorous grid capability studies conducted, before any wind generators are installed.

Part of the Biglow Canyon Wind Farm, Oregon, United States with a turbine under construction.

The wind blows faster at higher altitudes because of the reduced influence of drag. The increase in velocity with altitude is most dramatic near the surface and is affected by topography, surface roughness, and upwind obstacles such as trees or buildings. However, at higher altitudes, the power in the wind decreases proportional to the decrease in air density. Rendering significantly less efficient power extraction by the wind turbines, requiring for a higher investment for the same generation capacity at lower altitudes.

How closely to space the turbines together is a major factor in wind farm design. The closer the turbines are together the more the upwind turbines block wind from their rear neighbors (wake effect). However spacing turbines far apart increases the costs of roads and cables, and raises the amount of land needed to install a specific capacity of turbines. As a result of these factors, turbine spacing varies by site. Generally speaking manufacturers require 3.5 times the rotor diameter of the turbine between turbines as a minimum. Closer spacing is possible depending on the turbine model, the conditions at the site, and how the site will be operated. Airflows slow down as they approach an obstacle, known as the 'blockage effect', reducing available wind power by 2% for the turbines in front of other turbines.

Often in heavily saturated energy markets, the first step in site selection for large-scale wind projects before wind resource data collection is finding areas with adequate Available Transfer Capability (ATC). ATC is the measure of the remaining capacity in a transmission system available for further integration of generation without significant upgrades to transmission lines and substations, which have substantial costs, potentially undermining the viability of a project within that area, regardless of wind resource availability. Once a list of capable areas is constructed, the list is refined based on long term wind measurements, among other environmental or technical limiting factors such as proximity to load and land procurement.

Many Independent System Operators (ISO's) in the United States such as the California ISO and Midcontinent ISO use interconnection request queues to allow developers to propose new generation for a specific given area and grid interconnection. These request queues have both deposit costs at the time of request and ongoing costs for the studies the ISO will make for up to years after the request was submitted to ascertain the viability of the interconnection due to factors such as ATC. Larger corporations who can afford to bid the most queues will most likely have market power as to which sites with the most resource and opportunity get to be developed upon. After the deadline to request a place in the queue has passed, many firms will withdraw their requests after gauging the competition in order to make back some of the deposit for each request that is determined too risky in comparison to other larger firms' requests.

## Onshore

An aerial view of Whitelee Wind Farm, the largest onshore wind farm.

The world's first wind farm was 0.6 MW, consisting of 20 wind turbines rated at 30 kilowatts each, installed on the shoulder of Crotched Mountain in southern New Hampshire in December 1980.

Onshore turbine installations in hilly or mountainous regions tend to be on ridges generally three kilometres or more inland from the nearest shoreline. This is done to exploit the topographic acceleration as the wind accelerates over a ridge. The additional wind speeds gained in this way can increase energy produced because more wind goes through the turbines. The exact position of each turbine matters, because a difference of 30m could potentially double output. This careful placement is referred to as 'micro-siting'.

## Offshore

Offshore wind turbines.

Europe is the leader in offshore wind energy, with the first offshore wind farm (Vindeby) being installed in Denmark in 1991. As of 2010, there are 39 offshore wind farms in waters off Belgium, Denmark, Finland, Germany, Ireland, the Netherlands, Norway, Sweden and the United Kingdom, with a combined operating capacity of 2,396 MW. More than 100 GW (or 100,000 MW) of offshore projects are proposed or under development in Europe. The European Wind Energy Association has set a target of 40 GW installed by 2020 and 150 GW by 2030.

As of 2017, The Walney Wind Farm in the United Kingdom is the largest offshore wind farm in the world at 659 MW, followed by the London Array (630 MW) also in the UK.

Offshore wind turbines are less obtrusive than turbines on land, as their apparent size and noise is mitigated by distance. Because water has less surface roughness than land (especially deeper water), the average wind speed is usually considerably higher over open water. Capacity factors (utilisation rates) are considerably higher than for onshore locations.

The province of Ontario in Canada is pursuing several proposed locations in the Great Lakes, including the suspended Trillium Power Wind 1 approximately 20 km from shore and over 400 MW in size. Other Canadian projects include one on the Pacific west coast.

In 2010, there were no offshore wind farms in the United States, but projects were under development in wind-rich areas of the East Coast, Great Lakes, and Pacific coast; and in late 2016 the Block Island Wind Farm was commissioned.

Installation and service/maintenance of off-shore wind farms are a specific challenge for technology and economic operation of a wind farm. As of 2015, there are 20 jackup vessels for lifting components, but few can lift sizes above 5MW. Service vessels have to be operated nearly 24/7 (availability higher than 80% of time) to get sufficient amortisation from the wind turbines. Therefore, special fast service vehicles for installation (like Wind Turbine Shuttle) as well as for maintenance (including heave compensation and heave compensated working platforms to allow the service staff to enter the wind turbine also at difficult weather conditions) are required. So-called inertial and optical based Ship Stabilization and Motion Control systems (iSSMC) are used for that.

## Experimental and Proposed Wind Farms

Experimental wind farms consisting of a single wind turbine for testing purposes have been built. One such installation is Østerild Wind Turbine Test Field.

Airborne wind farms have been envisaged. Such wind farms are a group of airborne wind energy systems located close to each other connected to the grid at the same point.

Wind farms consisting of diverse wind turbines have been proposed in order to efficiently use wider ranges of wind speeds. Such wind farms are proposed to be projected under two criteria: maximization of the energy produced by the farm and minimization of its costs.

## By Region

## Australia

The Australian Canunda Wind Farm, South Australia at sunrise.

The Australian Greens have been significant supporters of Australian wind farms, however the party's former leader Bob Brown and the current leader Richard Di Natale have now both expressed concerns about environmental aspects of wind turbines, particularly the potential danger they impose for birds.

## China

Wind farm in Xinjiang, China.

In just five years, China leapfrogged the rest of the world in wind energy production, going from 2,599 MW of capacity in 2006 to 62,733 MW at the end of 2011. However, the rapid growth outpaced China's infrastructure and new construction slowed significantly in 2012.

At the end of 2009, wind power in China accounted for 25.1 gigawatts (GW) of electricity generating capacity, and China has identified wind power as a key growth component of the country's economy. With its large land mass and long coastline, China has exceptional wind resources. Researchers from Harvard and Tsinghua University have found that China could meet all of their electricity demands from wind power by 2030.

By the end of 2008, at least 15 Chinese companies were commercially producing wind turbines and several dozen more were producing components. Turbine sizes of 1.5 MW to 3 MW became common. Leading wind power companies in China were Goldwind, Dongfang Electric, and Sinovel along with most major foreign wind turbine manufacturers. China also increased production of small-scale wind turbines to about 80,000 turbines (80 MW) in 2008. Through all these developments, the Chinese wind industry appeared unaffected by the financial crisis of 2007–2008, according to industry observers.

According to the Global Wind Energy Council, the development of wind energy in China, in terms of scale and rhythm, is absolutely unparalleled in the world. The National People's Congress permanent committee passed a law that requires the Chinese energy companies to purchase all the electricity produced by the renewable energy sector.

## European Union

A wind farm in a mountainous area in Galicia, Spain.

The European Union has a total installed wind capacity of 93,957 MW. Germany has the third-largest capacity in the world (after China and the United States) with an installed capacity was 29,060 MW at the end of 2011, and Spain has 21,674 MW. Italy and France each had between 6,000 and 7,000 MW. By January 2014, the UK installed capacity was 10,495 MW. But energy production can be different from capacity – in 2010, Spain had the highest European wind power production with 43 TWh compared to Germany's 35 TWh.

Europe's largest windfarm is the 'London Array', an off-shore wind farm in the Thames Estuary in the United Kingdom, with a current capacity of 630 MW (the world's largest off-shore wind farm). Other large wind farms in Europe include Fântânele-Cogealac Wind Farm near Constanța, Romania with 600 MW capacity, and Whitelee Wind Farm near Glasgow, Scotland which has a total capacity of 539 MW.

An important limiting factor of wind power is variable power generated by wind farms. In most locations the wind blows only part of the time, which means that there has to be back-up capacity of conventional generating capacity to cover periods that the wind is not blowing. To address this issue it has been proposed to create a "supergrid" to connect national grids together across western Europe, ranging from Denmark across the southern North Sea to England and the Celtic Sea to Ireland, and further south to France and Spain especially in Higueruela which was for some time the biggest wind farm in the world. The idea is that by the time a low pressure area has moved away from Denmark to the Baltic Sea the next low appears off the coast of Ireland. Therefore, while it is true that the wind is not blowing everywhere all of the time, it will always be blowing somewhere.

## India

A wind farm overlooking Bada Bagh, India.

India has the fifth largest installed wind power capacity in the world. As of 31 March 2014, the installed capacity of wind power was 21136.3 MW mainly spread across Tamil Nadu state (7253 MW). Wind power accounts nearly 8.5% of India's total installed power generation capacity, and it generates 1.6% of the country's power.

## Jordan

The Tafila Wind Farm in Jordan, is the first large scale wind farm in the region.

The 117 MW Tafila Wind Farm in Jordan was inaugurated in December 2015, and is the first large scale wind farm project in the region.

## Morocco

Morocco has undertaken a vast wind energy program, to support the development of renewable energy and energy efficiency in the country. The Moroccan Integrated Wind Energy Project, spanning over a period of 10 years with a total investment estimated at $3.25 billion, will enable the country to bring the installed capacity, from wind energy, from 280 MW in 2010 to 2000 MW in 2020.

## Pakistan

Jhimpir Wind Farm, Pakistan

Pakistan has wind corridors in Jhimpir, Gharo and Keti Bundar in Sindh province and is currently developing wind power plants in Jhimpir and Mirpur Sakro (District Thatta). The government of Pakistan decided to develop wind power energy sources due to problems supplying energy to the southern coastal regions of Sindh and Balochistan. The Zorlu Energy Putin Power Plant is the first wind power plant in Pakistan. The wind farm is being developed in Jhimpir, by Zorlu Energy Pakistan the local subsidiary of a Turkish company. The total cost of project is $136 million. Completed in 2012, it has a total capacity of around 56MW. Fauji Fertilizer Company Energy Limited, has build a 49.5 MW wind Energy Farm at Jhimpir. Contract of supply of mechanical design was awarded to Nordex and Descon Engineering Limited. Nordex a German wind turbine manufacturer. In the end of 2011 49.6 MW will be completed.Pakistani Govt. also has issued LOI of 100 MW Wind power plant to FFCEL. Pakistani Govt. has plans to achieve electric power up to 2500 MW by the end of 2015 from wind energy to bring down energy shortage.

Currently four wind farms are operational (Fauji Fertilizer 49.5 MW (subsidiary of Fauji Foundation), Three Gorges 49.5 MW, Zorlu Energy Pakistan 56 MW, Sapphire Wind Power Co Ltd 52.6 MW) and six are under construction phase ( Master Wind Energy Ltd 52.6 MW, Sachal Energy Development Ltd 49.5 MW, Yunus Energy Ltd 49.5 MW, Gul Energy 49.5 MW, Metro Energy 49.5 MW, Tapal Energy ) and expected to achieve COD in 2017.

In Gharo wind corridor, two wind farms (Foundation Energy 1 & II each 49.5 MW) are operational while two wind farms Tenaga Generasi Ltd 49.5 MW and HydroChina Dawood Power Pvt Ltd 49.5 are under construction and expected to achieve COD in 2017.

According to a USAID report, Pakistan has the potential of producing 150,000 megawatts of wind energy, of which only the Sindh corridor can produce 40,000 megawatts.

## Philippines

The Philippines has the first windfarm in Southeast Asia. Located Northern part of the countries' biggest island Luzon, alongside the seashore of Bangui, Ilocos Norte.

The wind farm uses 20 units of 70-metre (230 ft) high Vestas V82 1.65 MW wind turbines, arranged on a single row stretching along a nine-kilometer shoreline off Bangui Bay, facing the West Philippine Sea.

Phase I of the NorthWind power project in Bangui Bay consists of 15 wind turbines, each capable of producing electricity up to a maximum capacity of 1.65 MW, for a total of 24.75 MW. The 15 on-shore turbines are spaced 326 metres (1,070 ft) apart, each 70 metres (230 ft) high, with 41 metres (135 ft) long blades, with a rotor diameter of 82 metres (269 ft) and a wind swept area of 5,281 square metres (56,840 sq ft). Phase II, was completed on August 2008, and added 5 more wind turbines with the same capacity, and brought the total capacity to 33 MW. All 20 turbines describes a graceful arc reflecting the shoreline of Bangui Bay, facing the West Philippine Sea.

Adjacent municipalities of Burgos and Pagudpud followed with 50 and 27 wind turbines with a capacity of 3 MW each for a Total of 150 MW and 81 MW respectively.

Two other wind farms were built outside of Ilocos Norte, the Pililla Wind Farm in Rizal and the Mindoro Wind Farm near Puerto Galera in Oriental Mindoro.

## Sri Lanka

Sri Lanka has received funding from the Asian Development Bank amounting to $300 million to invest in renewable energies. From this funding as well as $80 million from the Sri Lankan Government and $60 million from France's Agence Française de Développement, Sri Lanka is building two 100MW wind farms from 2017 due to be completed by late 2020 in Northern Sri Lanka.

## South Africa

Gouda Wind Facility, South Africa.

As of September 2015 a number of sizable wind farms have been constructed in South Africa mostly in the Western Cape region. These include the 100 MW Sere Wind Farm and the 138 MW Gouda Wind Facility.

Most future wind farms in South Africa are earmarked for locations along the Eastern Cape coastline. Eskom has constructed one small scale prototype windfarm at Klipheuwel in the Western Cape and another demonstrator site is near Darling with phase 1 completed. The first commercial wind farm, Coega Wind Farm in Port Elisabeth, was developed by the Belgian company Electrawinds.

## United States

San Gorgonio Pass wind farm, California.

U.S. wind power installed capacity in 2012 exceeded 51,630 MW and supplies 3% of the nation's electricity.

New installations place the U.S. on a trajectory to generate 20% of the nation's electricity by 2030 from wind energy. Growth in 2008 channeled some $17 billion into the economy, positioning wind power as one of the leading sources of new power generation in the country, along with natural gas. Wind projects completed in 2008 accounted for about 42% of the entire new power-producing capacity added in the U.S. during the year.

At the end of 2008, about 85,000 people were employed in the U.S. wind industry, and GE Energy was the largest domestic wind turbine manufacturer. Wind projects boosted local tax bases and revitalized the economy of rural communities by providing a steady income stream to farmers with wind turbines on their land. Wind power in the U.S. provides enough electricity to power the equivalent of nearly 9 million homes, avoiding the emissions of 57 million tons of carbon each year and reducing expected carbon emissions from the electricity sector by 2.5%.

Texas, with 10,929 MW of capacity, has the most installed wind power capacity of any U.S. state, followed by California with 4,570 MW and Iowa with 4,536 MW. The Alta Wind Energy Center (1,020 MW) in California is the nation's largest wind farm in terms of capacity. Altamont Pass Wind Farm is the largest wind farm in the U.S. in terms of the number of individual turbines.

## Criticism

## Environmental Impact

Compared to the environmental impact of traditional energy sources, the environmental impact of wind power is relatively minor. Wind power consumes no fuel, and emits no air pollution, unlike fossil fuel power sources. The energy consumed to manufacture and transport the materials used to build a wind power plant is equal to the new energy produced by the plant within a few months.

Onshore wind farms are criticized for their impact on the landscape. Their network of turbines, roads, transmission lines and substations can result in "energy sprawl". Typically they need to take up more land than other power stations and are more spread out. To power many major cities by wind alone would require building wind farms bigger than the cities themselves. Typically they also need to be built in wild and rural areas, which can lead to "industrialization of the country-side" and habitat loss. A report by the Mountaineering Council of Scotland concluded that wind farms have a negative impact on tourism in areas known for natural landscapes and panoramic views. However, land between the turbines can still be used for agriculture.

Livestock near a wind turbine.

Habitat loss and habitat fragmentation are the greatest impact of wind farms on wildlife. There are also reports of higher bird and bat mortality at wind turbines as there are around other artificial structures. The scale of the ecological impact may or may not be significant, depending on specific circumstances. The estimated number of bird deaths caused by wind turbines in the United States is between 140,000 and 328,000, whereas deaths caused by domestic cats in the United States are estimated to be between 1.3 and 4.0 billion birds each year and over 100 million birds are killed in the United States each year by impact with windows. Prevention and mitigation of wildlife fatalities, and protection of peat bogs, affect the siting and operation of wind turbines.

## Human Health

Wind turbines overlooking Ardrossan, Scotland.

There have been multiple scientific, peer-reviewed studies into wind farm noise, which have concluded that infrasound from wind farms is not a hazard to human health and there is no verifiable evidence for 'Wind Turbine Syndrome' causing Vibroacoustic disease, although some suggest further research might still be useful.

A 2007 report by the U.S. National Research Council noted that noise produced by wind turbines is generally not a major concern for humans beyond a half-mile or so. Low-frequency vibration and its effects on humans are not well understood and sensitivity to such vibration resulting from wind-turbine noise is highly variable among humans. There are opposing views on this subject, and more research needs to be done on the effects of low-frequency noise on humans.

In a 2009 report about "Rural Wind Farms", a Standing Committee of the Parliament of New South Wales, Australia, recommended a minimum setback of two kilometres between wind turbines and neighbouring houses (which can be waived by the affected neighbour) as a precautionary approach.

A 2014 paper suggests that the 'Wind Turbine Syndrome' is mainly caused by the nocebo effect and other psychological mechanisms. Australian science magazine Cosmos states that although the symptoms are real for those who suffer from the condition, doctors need to first eliminate known causes (such as pre-existing cancers or thyroid disease) before reaching definitive conclusions with the caveat that new technologies often bring new, previously unknown health risks.

## Effect on Power Grid

Utility-scale wind farms must have access to transmission lines to transport energy. The wind farm developer may be obliged to install extra equipment or control systems in the wind farm to meet the technical standards set by the operator of a transmission line.

The intermittent nature of wind power can pose complications for maintaining a stable power grid when wind farms provide a large percentage of electricity in any one region.

## Ground Radar Interference

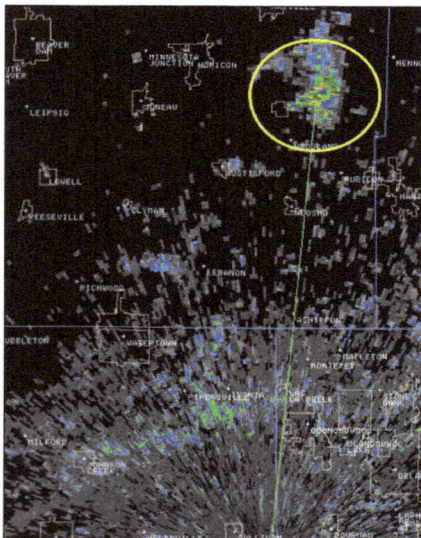

Wind farm interference (in yellow circle) on radar map.

Wind farms can interfere with ground radar systems used for military, weather and air traffic control. The large, rapidly moving blades of the turbines can return signals to the radar that can be mistaken as an aircraft or weather pattern. Actual aircraft and weather patterns around wind farms can be accurately detected, as there is no fundamental physical constraint preventing that. But aging radar infrastructure is significantly challenged with the task. The US military is using wind turbines on some bases, including Barstow near the radar test facility.

## Effects

The level of interference is a function of the signal processors used within the radar, the speed of the aircraft and the relative orientation of wind turbines/aircraft with respect to the radar. An aircraft flying above the wind farm's turning blades could become impossible to detect because the blade tips can be moving at nearly aircraft velocity. Studies are currently being performed to determine the level of this interference and will be used in future site planning. Issues include masking (shadowing), clutter (noise), and signal alteration. Radar issues have stalled as much as 10,000 MW of projects in USA.

Some very long range radars are not affected by wind farms.

## Mitigation

Permanent problem solving include a non-initiation window to hide the turbines while still tracking aircraft over the wind farm, and a similar method mitigates the false returns. England's Newcastle Airport is using a short-term mitigation; to blank the turbines on the radar map with a software patch. Wind turbine blades using stealth technology are being developed to mitigate radar reflection problems for aviation. As well as stealth windfarms, the future development of infill radar systems could filter out the turbine interference.

A mobile radar system, the Lockheed Martin TPS-77, can distinguish between aircraft and wind turbines, and more than 170 TPS-77 radars are in use around the world.

## Radio Reception Interference

There are also reports of negative effects on radio and television reception in wind farm communities. Potential solutions include predictive interference modelling as a component of site selection.

Wind turbines can often cause terrestrial television interference when the direct path between television transmitter and receiver is blocked by terrain. Interference effects become significant when the reflected signal from the turbine blades approaches the strength of the direct unreflected signal. Reflected signals from the turbine blades can cause loss of picture, pixellation and disrupted sound. There is a common misunderstanding that digital TV signals will not be affected by turbines — in practice they are.

## Agriculture

A 2010 study found that in the immediate vicinity of wind farms, the climate is cooler during the day and slightly warmer during the night than the surrounding areas due to the turbulence generated by the blades.

In another study an analysis carried out on corn and soybean crops in the central areas of the United States noted that the microclimate generated by wind turbines improves crops as it prevents the late spring and early autumn frosts, and also reduces the action of pathogenic fungi that grow on the leaves. Even at the height of summer heat, the lowering of 2.5–3 degrees above the crops due to turbulence caused by the blades, can make a difference for the cultivation of corn.

## Offshore Wind Energy

Wind turbines and electrical substation of
Alpha Ventus Offshore Wind Farm in the North Sea.

Offshore wind power or offshore wind energy is the use of wind farms constructed in bodies of water, usually in the ocean on the continental shelf, to harvest wind energy to generate electricity. Higher wind speeds are available offshore compared to on land, so offshore wind power's electricity generation is higher per amount of capacity installed, and NIMBY opposition to construction is usually much weaker. Unlike the typical use of the term "offshore" in the marine industry, offshore wind power includes inshore water areas such as lakes, fjords and sheltered coastal areas, using traditional fixed-bottom wind turbine technologies, as well as deeper-water areas using floating wind turbines.

At the end of 2017, the total worldwide offshore wind power capacity was 18.8 gigawatt (GW). All the largest offshore wind farms are currently in northern Europe, especially in the United Kingdom and Germany, which together account for over two-thirds of the total offshore wind power installed worldwide. As of September 2018, the 659 MW Walney Extension in the United Kingdom is the largest offshore wind farm in the world. The Hornsea Wind Farm under construction in the United Kingdom will become the largest when completed, at 1,200 MW. Other projects are in the planning stage, including Dogger Bank in the United Kingdom at 4.8 GW, and Greater Changhua in Taiwan at 2.4 GW.

The cost of offshore wind power has historically been higher than that of onshore wind generation, but costs have been decreasing rapidly in recent years to $78/MWh in 2019. Offshore wind power in Europe has been price-competitive with conventional power sources since 2017.

### Future Development

Projections for 2020 estimate an offshore wind farm capacity of 40 GW in European waters, which would provide 4% of the European Union's demand of electricity. The European Wind Energy Association has set a target of 40 GW installed by 2020 and 150 GW by 2030. Offshore wind power capacity is expected to reach a total of 75 GW worldwide by 2020, with significant contributions from China and the United States.

The Organisation for Economic Co-operation and Development (OECD) predicted in 2016 that offshore wind power will grow to 8% of ocean economy by 2030, and that its industry will employ 435,000 people, adding $230 billion of value.

## Types of Offshore Wind Turbines

## Fixed Foundation Offshore Wind Turbines

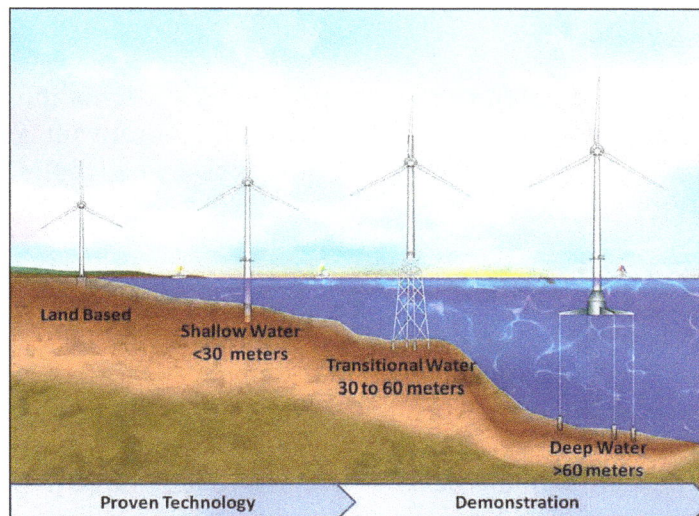

Progression of expected wind turbine evolution to deeper water.

Almost all currently operating offshore wind farms employ fixed foundation turbines, with the exception of a few pilot projects. Fixed foundation offshore wind turbines have fixed foundations underwater, and are installed in relatively shallow waters of up to 50–60 m.

Types of underwater structures include monopile, tripod, and jacketed, with various foundations at the sea floor including monopile or multiple piles, gravity base, and caissons. Offshore turbines require different types of bases for stability, according to the depth of water. To date a number of different solutions exist:

- A monopile (single column) base, six meters in diameter, is used in waters up to 30 meters deep.

- Gravity base structures, for use at exposed sites in water 20–80 m deep.

- Tripod piled structures, in water 20–80 m deep.

- Tripod suction caisson structures, in water 20–80 m deep.

- Conventional steel jacket structures, as used in the oil and gas industry, in water 20–80 m deep.

Monopiles up to 11 m diameter at 2,000 tonnes can be made, but the largest so far are 1,300 tons which is below the 1,500 tonnes limit of some crane vessels. The other turbine components are much smaller.

The tripod pile substructure system is a new concept developed to reach deeper waters than with the shallow water systems, up to 60 m. This technology consists of three monopiles linked together through a joint piece at the top. The main advantage of this solution is the simplicity of the installation, which is done by installing three monopiles and then adding the upper joint.

Tripod is an innovative concept that consists on a central pipe that lies on a tripod tubular frame configuration at its bottom part. This uses three small seabed driven piles at each leg of the tripod to link it to the seabed. The main advantage of the tripod system is that it has a larger base, which decreases its risk of getting overturned. Due to the large dimensions the installation process is more difficult and increases the cost.

A steel jacket structure comes from an adaptation to the offshore wind industry of concepts that have been in use in the oil and gas industry for decades. Their main advantage lies in the possibility of reaching higher depths (up to 80m). Their main limitations are due to the high construction and installation costs.

## Floating Offshore Wind Turbines

For locations with depths over about 60–80 m, fixed foundations are uneconomical or technically unfeasible, and floating wind turbine anchored to the ocean floor are needed. Hywind is the world's first full-scale floating wind turbine, installed in the North Sea off Norway in 2009. Hywind Scotland, commissioned in October 2017, is the first operational floating wind farm, with a capacity of 30 MW. Other kinds of floating turbines have been deployed, and more projects are planned.

## Vertical Axis Offshore Wind Turbines

Although the great majority of onshore and all large-scale offshore wind turbines currently installed are horizontal axis, vertical axis wind turbines have been proposed for use in offshore installations. Thanks to the installation offshore and their lower center of gravity, these turbines can in principle be built bigger than horizontal axis turbines, with proposed designs of up to 20 MW capacity per turbine. This could improve the economy of scale of offshore wind farms. However, there are no current large-scale demonstrations of this technology.

## Economics

The advantage of locating wind turbines offshore is that the wind is much stronger off the coasts, and unlike wind over the continent, offshore breezes can be strong in the afternoon, matching the time when people are using the most electricity. Offshore turbines can also be located close to the load centers along the coasts, such as large cities, eliminating the need for new long-distance transmission lines. However, there are several disadvantages of offshore installations, related to more expensive installation, difficulty of access, and harsher conditions for the units.

Locating wind turbines offshore exposes the units to high humidity, salt water and salt water spray which negatively affect service life, cause corrosion and oxidation, increase maintenance and repair costs and in general make every aspect of installation and operation much more difficult, time-consuming, more dangerous and far more expensive than sites on land. The humidity and temperature is controlled by air conditioning the sealed nacelle. Sustained high-speed operation and generation also increases wear, maintenance and repair requirements proportionally.

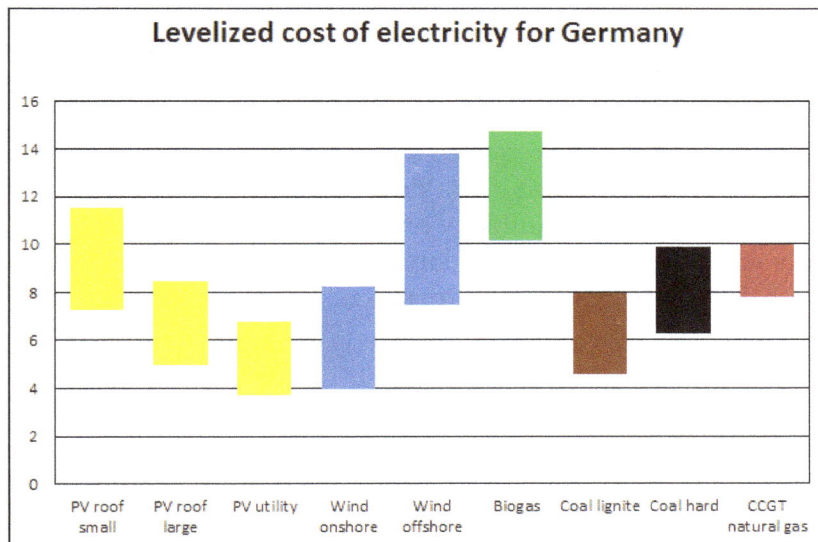

Comparison of the levelized cost of electricity of offshore
wind power compared to other sources in Germany.

The cost of the turbine represents just one third to one half of total costs in offshore projects today, the rest comes from infrastructure, maintenance, and oversight. Costs for foundations, installation, electrical connections and operation and maintenance (O&M) are a large share of the total for offshore installations compared to onshore wind farms. The cost of installation and electrical connection also increases rapidly with distance from shore and water depth.

Other limitations of offshore wind power are related to the still limited number of installations. The offshore wind industry is not yet fully industrialized, as supply bottlenecks still exist as of 2017.

## Investment Costs

Offshore wind farms tend to have larger turbines when compared to onshore installations, and the trend is towards a continued increase in size. Economics of offshore wind farms tend to favor larger turbines, as installation and grid connection costs decrease per unit energy produced. Moreover, offshore wind farms do not have the same restriction in size of onshore wind turbines, such as availability of land or transportation requirements.

## Operating Costs

Operational expenditures for wind farms are split between Maintenance (38%), Port Activities (31%), Operation (15%), License Fees (12%), and Miscellaneous Costs (4%).

Operation and maintenance costs typically represent 53% of operational expenditures, and 25% - 30% of the total lifecycle costs for offshore wind farms. O&Ms are considered one of the major barriers for further development of this resource.

Maintenance of offshore wind farms is much more expensive than for onshore installations. For example, a single technician in a pickup truck can quickly, easily and safely access turbines on land in almost any weather conditions, exit his or her vehicle and simply walk over to and into the turbine tower to gain access to the entire unit within minutes of arriving onsite. Similar access to offshore turbines involves driving to a dock or pier, loading necessary tools and supplies into boat, a voyage to the wind turbine(s), securing the boat to the turbine structure, transferring tools and supplies to and from boat to turbine and turbine to boat and performing the rest of the steps in reverse order. In addition to standard safety gear such as a hardhat, gloves and safety glasses, an offshore turbine technician may be required to wear a life vest, waterproof or water-resistant clothing and perhaps even a survival suit if working, sea and atmospheric conditions make rapid rescue in case of a fall into the water unlikely or impossible. Typically at least two technicians skilled and trained in operating and handling large power boats at sea are required for tasks that one technician with a driver's license can perform on land in a fraction of the time at a fraction of the cost.

## Cost of Energy

Cost for installed offshore turbines fell 30% to $78/MWh in 2019, a more rapid drop than other types of renewable energy. It has been suggested that innovation at scale could deliver 25% cost reduction in offshore wind by 2020. Offshore wind power market plays an important role in achieving the renewable target in most of the countries around the world.

Auctions in 2016 for future projects have reached costs of €54.5 per megawatt hour (MWh) at the 700 MW Borssele 3&4 due to government tender and size, and €49.90 per MWh (without transmission) at the 600 MW Kriegers Flak.

In September 2017 contracts were awarded in the United Kingdom for a strike price of £57.50 per MWh making the price cheaper than nuclear and competitive with gas.

In September 2018 contracts were awarded for Vineyard Wind, Massachusetts, USA at a cost of between $65-$74 per MWh.

## Offshore Wind Resources

Offshore wind resource characteristics span a range of spatial and temporal scales and field data on external conditions. For the North Sea, wind turbine energy is around 30 kWh/m$^2$ of sea area, per year, delivered to grid. The energy per sea area is roughly independent of turbine size.

## Planning and Permitting

A number of things are necessary in order to attain the necessary information for planning the commissioning of an offshore wind farm. The first information required is offshore wind characteristics. Additional necessary data for planning includes water depth, currents, seabed, migration, and wave action, all of which drive mechanical and structural loading on potential turbine configurations. Other factors include marine growth, salinity, icing, and the geotechnical characteristics of the sea or lake bed.

Existing hardware for measurements includes Light Detection and Ranging (LIDAR), Sonic Detection and Ranging (SODAR), radar, autonomous underwater vehicles (AUV), and remote satellite sensing, although these technologies should be assessed and refined, according to a report from a coalition of researchers from universities, industry, and government, supported by the Atkinson Center for a Sustainable Future.

Because of the many factors involved, one of the biggest difficulties with offshore wind farms is the ability to predict loads. Analysis must account for the dynamic coupling between translational (surge, sway, and heave) and rotational (roll, pitch, and yaw) platform motions and turbine motions, as well as the dynamic characterization of mooring lines for floating systems. Foundations and substructures make up a large fraction of offshore wind systems, and must take into account every single one of these factors. Load transfer in the grout between tower and foundation may stress the grout, and elastomeric bearings are used in several British sea turbines.

Corrosion is also a serious problem and requires detailed design considerations. The prospect of remote monitoring of corrosion looks very promising using expertise utilised by the offshore oil/ gas industry and other large industrial plants.

Some of the guidelines for designing offshore wind farms are IEC 61400-3, but in the US several other standards are necessary. In the EU, different national standards are to be straightlined into more cohesive guidelines to lower costs. The standards requires that a loads analysis is based on site-specific external conditions such as wind, wave and currents.

The planning and permitting phase can cost more than $10 million, take 5–7 years and have an uncertain outcome. The industry puts pressure on the governments to improve the processes. In Denmark, many of these phases have been deliberately streamlined by authorities in order to minimize hurdles, and this policy has been extended for coastal wind farms with a concept called 'one-stop-shop'. The United States introduced a similar model called Smart from the Start in 2012.

## Turbine Construction Materials Considerations

Since offshore wind turbines are located in oceans and large lakes, the materials used for the turbines have to be modified from the materials used for land based wind turbines and optimized for corrosion resistance to salt water and the new loading forces experienced by the tower being partially submerged in water. With one of the main reasons for interest in offshore wind power being the higher wind speeds, some of the loading differences will come from higher shearing forces between the top and bottom of the wind turbine due to differences in wind speeds. There should also be considerations for the buffeting loads that will be experienced by the waves around the base of the tower, which converges on the use of steel tubular towers for offshore wind applications.

Since there's constant exposure to salt and water for offshore wind turbines, the steel used for the monopile and turbine tower must be treated for corrosion resistance, especially at the base of the tower in the "splash zone" for waves breaking against the tower and in the monopile. Two techniques that can be used include cathode protection and the use of coatings to reduce corrosion pitting, which is a common source for hydrogen induced stress cracking. For cathode protection, galvanized anodes are attached to the monopile and have enough of a potential difference with the steel to be preferentially corroded over the steel used in the monopile. Some coatings that have been applied to offshore wind turbines include hot dip zinc coatings and 2-3 epoxy coatings with a polyurethane topcoat.

## Installation

Several foundation structures for offshore wind turbines in a port.

Specialized jackup rigs (Turbine Installation Vessels) are used to install foundation and turbine. As of 2019 the next generation of vessels are being built, capable of lifting 3-5,000 tons to 160 meters.

A large number of monopile foundations have been used in recent years for economically constructing fixed-bottom offshore wind farms in shallow-water locations. Each uses a single, generally large-diameter, foundation structural element to support all the loads (weight, wind, etc.) of a large above-surface structure.

The typical construction process for a wind turbine sub-sea monopile foundation in sand includes using a pile driver to drive a large hollow steel pile 25 m deep into the seabed, through a 0.5 m layer of larger stone and gravel to minimize erosion around the pile. These piles can be 4 m in diameter with approximately 50mm thick walls. A transition piece (complete with pre-installed features such as boat-landing arrangement, cathodic protection, cable ducts for sub-marine cables, turbine tower flange, etc.) is attached to the now deeply driven pile, the sand and water are removed from the centre of the pile and replaced with concrete. An additional layer of even larger stone, up to 0.5 m diameter, is applied to the surface of the seabed for longer-term erosion protection.

For the ease of installing the towers and connecting them to the seabed, they are installed in two parts, the portion below the water surface and the portion above the water. The two portions of the tower are joined by a transition piece which is filled with a grouted connection. The grouted connected helps transfer the loads experienced by the turbine tower to the more stable monopile foundation of the turbine. One technique for strengthening the grout used in the connections is to include weld beads known as shear keys along the length of the grout connection to prevent any sliding between the monopile and the tower.

## Grid Connection

There are several different types of technologies that are being explored as viable options for integrating offshore wind power into the onshore grid. The most conventional method is through high-voltage alternating current (HVAC) transmission lines. HVAC transmission lines are currently the most commonly used form of grid connections for offshore wind turbines. However, there are significant limitations that prevent HVAC from being practical, especially as the distance to offshore turbines increases. First, HVAC is limited by cable charging currents, which are a result of capacitance in the cables. Undersea AC cables have a much higher capacitance than overhead AC cables, so losses due to capacitance become much more significant, and the voltage magnitude at the receiving end of the transmission line can be significantly different from the magnitude at the receiving end. In order to compensate for these losses, either more cables or reactive compensation must be added to the system. Both of these add costs to the system. Additionally, because HVAC cables have both real and reactive power flowing through them, there can be additional losses. Because of these losses, underground HVAC lines are limited in how far they can extend. The maximum appropriate distance for HVAC transmission for offshore wind power is considered to be around 80 km.

An offshore structure for housing an HVDC converter station for offshore wind parks is being moved by a heavy-lift ship in Norway.

Using high-voltage direct current (HVDC) cables has been a proposed alternative to using HVAC cables. HVDC transmission cables are not affected by the cable charging currents and experience less power loss because HVDC does not transmit reactive power. With less losses, undersea HVDC lines can extend much farther than HVAC. This makes HVDC preferable for siting wind turbines very far offshore. However, HVDC requires power converters in order to connect to the AC grid. Both line commutated converters (LCCs) and voltage source converters (VSCs) have been considered for this. Although LCCs are a much more widespread technology and cheaper, VSCs have many more benefits, including independent active power and reactive power control. New research has been put into developing hybrid HVDC technologies that have a LCC connected to a VSC through a DC cable.

In order to transport the energy from offshore wind turbines to onshore energy plants, cabling has to be placed along the ocean floor. The cabling has to be able to transfer large amounts of current efficiently which requires optimization of the materials used for the cabling as well as determining cable paths for the use of a minimal amount of cable materials. One way to reduce the cost of the cables used in these applications is to convert the copper conductors to aluminum conductors, however the suggested replacement brings up an issue of increased cable motion and potential damage since aluminum is less dense than copper.

## Maintenance

Turbines are much less accessible when offshore (requiring the use of a service vessel or helicopter for routine access, and a jackup rig for heavy service such as gearbox replacement), and thus reliability is more important than for an onshore turbine. Some wind farms located far from possible onshore bases have service teams living on site in offshore accommodation units. To limit the effects of corrosion on the blades of a wind turbine, a protective tape of elastomeric materials is applied, though the droplet erosion protection coatings provide better protection from the elements.

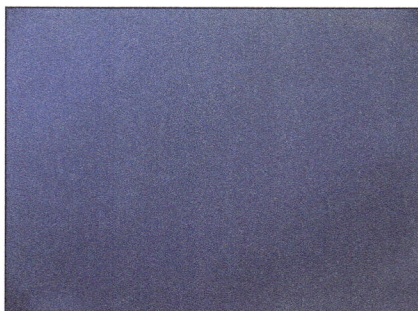

Offshore wind turbines of the Rødsand Wind Farm in the Fehmarn Belt,
the western part of the Baltic Sea between Germany and Denmark.

A maintenance organization performs maintenance and repairs of the components, spending almost all its resources on the turbines. The conventional way of inspecting the blades is for workers to rappel down the blade, taking a day per turbine. Some farms inspect the blades of three turbines per day by photographing them from the monopile through a 600mm lens, avoiding to go up. Others use camera drones.

Because of their remote nature, prognosis and health-monitoring systems on offshore wind turbines will become much more necessary. They would enable better planning just-in-time maintenance, thereby reducing the operations and maintenance costs. According to a report from a coalition of researchers from universities, industry, and government (supported by the Atkinson Center for a Sustainable Future), making field data from these turbines available would be invaluable in validating complex analysis codes used for turbine design. Reducing this barrier would contribute to the education of engineers specializing in wind energy.

## Decommissioning

As the first offshore wind farms reach their end of life, a demolition industry develops to recycle them at a cost of DKK 2-4 million per MW, to be guaranteed by the owner. The first offshore wind farm to be decommissioned was Yttre Stengrund in Sweden in November 2015, followed by Vindeby in 2017 and Blyth in 2019.

## Environmental Impact

Offshore wind farms have very low global warming potential per unit of electricity generated, comparable to that of onshore wind farms. Offshore installations also have the advantage of limited impact of noise and on the landscape compared to land-based projects.

While the offshore wind industry has grown dramatically over the last several decades, there is still a great deal of uncertainty associated with how the construction and operation of these wind farms affect marine animals and the marine environment. Common environmental concerns associated with offshore wind developments include:

- The risk of seabirds being struck by wind turbine blades or being displaced from critical habitats.

- The underwater noise associated with the installation process of driving monopile turbines into the seabed.

- The physical presence of offshore wind farms altering the behavior of marine mammals, fish, and seabirds with attraction or avoidance.

- The potential disruption of the nearfield and farfield marine environment from large offshore wind projects.

- The Tethys database provides access to scientific literature and general information on the potential environmental effects of offshore wind energy.

## Onshore Wind Energy

Wind turbines harness the energy of moving air to generate electricity. Onshore wind refers to turbines located on land, while offshore turbines are located out at sea or in freshwater. In the UK, the pros and cons of onshore wind energy, in comparison with other low-carbon and fossil fuel energy sources, have recently been the subject of debate in the press and among politicians.

Onshore wind already plays a leading role in the generation of renewable electricity in the UK. In 2010, it generated around 7TWh – more than a quarter of the electricity provided by British renewables at that time and enough to save six million tonnes of $CO_2$, according to government estimates. By 2020, onshore wind is expected to generate up to 30TWh. Onshore wind can therefore play a key role in reducing greenhouse gas emissions created by the UK's power sector – which will be crucial to meeting the UK's legally binding carbon budgets.

Onshore wind has the advantage of being one of the most affordable renewable energy sources. Generating electricity from onshore wind turbines typically costs around 7–9p per kWh, which is around half the cost of offshore wind and a quarter of the costs of solar photovoltaic panels. It is also slightly cheaper, on average, than nuclear power. Onshore wind generation is still slightly more expensive than fossil fuels (generating electricity from gas power plants currently costs between 4.1 and 7.5 p/kWh), but its price is expected to fall in the coming years.

Some emissions are created by the manufacture, transportation and installation of wind turbines, but these are considered fairly low. Additional emissions are attributed to the fact that wind energy (like solar and wave power) is intermittent, generating electricity only when the wind is blowing, and at sufficient strength. When wind strength is insufficient for turbines to operate, fossil-fuel-based power supply is needed as "backup". The current small proportion of renewable electricity in the UK market requires very little backup, but as the share increases additional backup will be needed. However, other technologies, such as inter-linkages with other countries' grids, energy storage and electricity demand management, are expected to help tackle intermittency in the future, so the overall future impact on emissions is considered relatively low.

Onshore wind has been criticised for its visual impact. Although other power infrastructure, like fossil fuel and nuclear power stations, can also modify landscapes and habitats, onshore wind turbines are typically more spread out than other large-scale energy infrastructure projects and so can affect a larger area. Another criticism is that species such as birds and bats may also be affected by wind turbines – though bird fatalities due to turbine collisions are relatively low compared to other fatality causes, such as traffic and domestic cats. Impacts on wildlife can be minimised by careful site selection and by avoiding areas of high conservation or habitat value. A third potential issue is that turbines can contribute to noise pollution, but government studies find noise levels are comparatively low and should not significantly impact on nearby residents. New guidance is being drafted to inform future planning policy on noise issues.

Environmental Impact Assessments review these kinds of potential impacts on a case-by-case basis and seek to protect unsuitable areas, such as those of high conservation or heritage value. In some cases, undesirable local impacts may make more expensive renewable technologies, such as offshore wind or solar, more attractive. The extra cost of offshore wind can be seen as the premium society is willing to pay in order to avoid the local environmental cost of onshore turbines.

The choice between more affordable electricity (which would favour onshore wind) and local environmental protection (which may favour other low-carbon technologies) is ultimately a societal and political one. Given the economic and environmental trade-offs, technological uncertainty and the absence of one clear winner when it comes to energy sources, many economists suggest (pdf) the best approach is a portfolio of different technologies to balance the cost to consumers and environmental concerns.

## Advantages and Disadvantages of Offshore Wind Energy and Onshore Wind Power

Offshore Wind Energy has been foretasted from around 3 GW to 75 GW by 2020 as countries in Europe, Asia and North America heavily support this industry. Note onshore wind energy growth has slowed down as 38 GW of Wind Capacity were installed in 2010.Offshore Wind Energy is being subsidized as it has a number of advantages over onshore wind power despite its higher costs. Top Wind Turbine companies like Gamesa, Vestas, Siemens are jostling to get a first mover advantage as the learning curve is steep. The opportunity is so large that South Korea shipbuilders like Samsung, Daewoo and Hyundai are investing heavily in this area. Note Onshore Wind Energy also has many advantages which have overcome the drawbacks to become the biggest green industry. Offshore Wind Power is set to take up the growth baton for Onshore Wind as Costs of Offshore Wind are expected to come down as the industry and technology matures and more Wind Turbine Companies gain experience in this area. Note USA has 4000 GW of Offshore Wind Capacity which is enough to meet the entire electricity need of the biggest power consumption country in the world.

## Advantages of Offshore Wind Farms

Wind Energy has a number of advantages and a small number of drawbacks. Offshore Wind Farms manage to overcome even these few drawbacks.

## Advantages Common to Offshore and Onshore Wind Energy

- No Pollution and Global Warming Effects – Wind Turbines does not lead to pollution which is one of the biggest advantages of Wind Energy. Note there are costs associated with the equipment used to build and transport Wind Equipment but the running of Wind Energy leads to no pollution.

- Low Costs – The Costs of Wind Energy has reached the level of Gas powered Energy and can be generated at extremely low rates of around 7-8c/KwH in favorable conditions.

- Big Industrial Base – Wind Energy has become a mainstream source of energy and a large industrial base already exists.This allows a rapid deployment of Wind Power in most places in the world. The number of Wind Turbine Producers is increasing with a number of Asian firms entering the industry.

- No Fuel Cost – Wind Energy does not require any fuel like most other sources of renewable energy. This is a huge advantage over other fossil fuels whose costs are increasing at a drastic rate every year. Electricity prices are increasingly rapidly in most parts of the world much faster than general inflation. Price shocks due to high fuel costs are a big risk with fossil fuel energy these days.

## Advantages of Offshore Wind over Onshore Wind

- No Noise Pollution – Wind Turbines emit a slight whirring noise which has led to problems with people living nearby. Some farmers have also complained that the livestock like sheep get affected by the moving of the Wind Blades. Offshore Wind Farms are located far off the coast cause no such noise problems for humans or wildlife.

- No Injuries to Birds – Older Wind Farms on Land frequently cause deaths and injuries to birds though newer wind turbines don't cause too much problems. Offshore Wind Farms do away with this problem entirely as they are located in the Ocean where birds don't fly frequently if at all. There is research being conducted to see if there is an impact on sea life by Cowries.

- No loss in scenery though near shore offshore wind farms have come into controversy because of this, the Cape Wind Project is attracting a lot of protests particularly from the Kennedy's who say that it will destroy the view from their house near the ocean.

## Disadvantages of Offshore Wind over Onshore Wind

Cost – This is the biggest disadvantage of offshore wind power over onshore wind energy. Note it can cost between 2.5-3.5 times more to generate electricity from offshore wind turbines than the wind farms built on land. There are a number of factors that determine the price such as wind speeds etc. However offshore wind industry is still in a novice state compared to the relatively mature level of the land based wind industry.

## Advantages and Challenges of Wind Energy

Wind energy offers many advantages, which explains why it's one of the fastest-growing energy sources in the world. Research efforts are aimed at addressing the challenges to greater use of wind energy.

### Advantages of Wind Power

- Wind power is cost-effective. Land-based utility-scale wind is one of the lowest-priced energy sources available today, costing between two and six cents per kilowatt-hour, depending on the wind resource and the particular project's financing. Because the electricity from wind farms is sold at a fixed price over a long period of time (e.g. 20+ years) and its fuel is free, wind energy mitigates the price uncertainty that fuel costs add to traditional sources of energy.

- Wind creates jobs. The U.S. wind sector employed more than 100,000 workers in 2016, and wind turbine technician is one of the fastest-growing American jobs of the decade. According to the Wind Vision Report, wind has the potential to support more than 600,000 jobs in manufacturing, installation, maintenance, and supporting services by 2050.

- Wind enables U.S. industry growth and U.S. competitiveness. Wind has an annual economic impact of about $20 billion on the U.S. economy, The United States has a vast domestic resources and a highly-skilled workforce, and can compete globally in the clean energy economy.

- It's a clean fuel source. Wind energy doesn't pollute the air like power plants that rely on combustion of fossil fuels, such as coal or natural gas, which emit particulate matter, nitrogen oxides, and sulfur dioxide—causing human health problems and economic damages. Wind turbines don't produce atmospheric emissions that cause acid rain, smog, or greenhouse gases.

- Wind is a domestic source of energy. The nation's wind supply is abundant and inexhaustible. Over the past 10 years, cumulative wind power capacity in the United States increased an average of 30% per year, and wind now has the largest renewable generation capacity of all renewables in the United States.

- It's sustainable. Wind is actually a form of solar energy. Winds are caused by the heating of the atmosphere by the sun, the rotation of the Earth, and the Earth's surface irregularities. For as long as the sun shines and the wind blows, the energy produced can be harnessed to send power across the grid.

- Wind turbines can be built on existing farms or ranches. This greatly benefits the economy in rural areas, where most of the best wind sites are found. Farmers and ranchers can continue to work the land because the wind turbines use only a fraction of the land. Wind power plant owners make rent payments to the farmer or rancher for the use of the land, providing landowners with additional income.

## Challenges of Wind Power

Wind power must still compete with conventional generation sources on a cost basis. Depending on how energetic a wind site is, the wind farm might not be cost competitive. Even though the cost of wind power has decreased dramatically in the past 10 years, the technology requires a higher initial investment than fossil-fueled generators.

Good wind sites are often located in remote locations, far from cities where the electricity is needed. Transmission lines must be built to bring the electricity from the wind farm to the city. However, building just a few already-proposed transmission lines could significantly reduce the costs of expanding wind energy.

Wind resource development might not be the most profitable use of the land. Land suitable for wind-turbine installation must compete with alternative uses for the land, which might be more highly valued than electricity generation.

Turbines might cause noise and aesthetic pollution. Although wind power plants have relatively little impact on the environment compared to conventional power plants, concern exists over the noise produced by the turbine blades and visual impacts to the landscape.

Turbine blades could damage local wildlife. Birds have been killed by flying into spinning turbine blades. Most of these problems have been resolved or greatly reduced through technological development or by properly siting wind plants.

## References

- Wind, science: britannica.com, Retrieved 14 May, 2019

- Walwyn, David Richard; Brent, Alan Colin (2015). "Renewable energy gathers steam in South Africa". Renewable and Sustainable Energy Reviews. 41: 390. Doi:10.1016/j.rser.2014.08.049. Hdl:2263/49731

- Keynotes-on-5-different-types-of-wind, notes: yourarticlelibrary.com, Retrieved 15 June, 2019

- Kart, Jeff (13 May 2009). "Wind, Solar-Powered Street Lights Only Need a Charge Once Every Four Days". Clean Technica. Clean Technica. Retrieved 30 April 2012

- Climate-change-windpower, environment: theguardian.com, Retrieved 16 July, 2019

- Anaya-Lara, Olimpo; Campos-Gaona, David; Moreno-Goytia, Edgar; Adam, Grain (10 April 2014). Grid Integration of Offshore Wind Farms – Case Studies. Wiley. Doi:10.1002/9781118701638.ch5. ISBN 9781118701638

- Offshore-wind-energy-vs-onshore-wind-power-advantages-and-disadvantages: greenworldinvestor.com, Retrieved 17 August, 2019

- Tillessen, Teena (2010). "High demand for wind farm installation vessels". Hansa International Maritime Journal. Vol. 147 no. 8. Pp. 170–17

- Advantages-and-challenges-wind-energy, wind, eere: energy.gov, Retrieved 18 January, 2019

# Types of Wind Turbines

# 2

- **Working Principle of Wind Turbine**
- **Small Wind Turbines**
- **Applications of Small Wind Turbines**
- **Wind-turbine Aerodynamics**
- **Variable Speed Wind Turbine**
- **Types of Wind Turbine**
- **Darrieus Wind Turbine**
- **Savonius Wind Turbine**

Wind turbines are the devices used for the conversion of kinetic energy of wind into electrical energy. The two main types of wind turbines are horizontal-axis turbines and vertical-axis turbines. This is an introductory chapter which will briefly introduce these types of wind turbines.

Wind turbine operate by using the kinetic energy of the wind, which pushes the blades of the turbine and spins a motor that converts the kinetic energy into electrical energy for consumer use.

Wind Turbines are rotating machines that can be used directly for grinding or can be used to generate electricity from the kinetic power of the wind. They provide the clean and renewable energy for us of both home and office. Wind Turbines are a great way to save money and make the environment clean and green.

This process has been adapted for use for various applications and can be seen in use by boats, traffic signs, or whole communities that use a wind farm for power. The development of wind turbines is a major step towards overhauling the way we produce our energy.

Basically there are two types of wind generators, those with vertical axis and those with horizontal axis. They can be used to generate electricity both onshore and offshore. Wind Turbines can be combined to form clusters called "wind farms" which are used by large companies to use that power as their backup. Apart from generating electricity they can also be used for grain-grinding, water pumping, charging batteries.

Historically, wind turbines were used for sailing, irrigation and grinding-grains. It was in the early 20th century that it was used for generating power. Today, large wind turbines can be seen in the rural areas or near the sea coast where speed of the wind is generally throughout the day. Device called wind resource assessment is used for estimating the wind speed.

## Wind Turbine Components

Wind turbine systems are made up of many different pieces of equipment that all serve a purpose in delivering electricity where it is intended to go. Of those many different pieces, the below list serves as a general blueprint for the main components that can and often times will be found in wind turbine systems regardless of the type of design.

- Rotor – The rotor is made of blades that are attached to a center piece. The blades are shaped such that when the wind pushes against them they turn.

- Pitch Drive – Used to rotate blades to accommodate for high-speed wind.

- Nacelle – The rotor is attached to a housing unit called a nacelle, which protects various other components necessary to the wind turbine operation.

- Brake – Necessary to slow the rotor down.

- Low-Speed Shaft – Attaches to the rotor and turns as the rotor turns on a 1:1 ratio.

- Gear Box – Serves the same function as a car, the rotor spins slowly as the wind pushes against it and the gearbox or transmission increases that rotational speed for the generator.

- High-Speed Shaft – Attaches to the gearbox and generator and spins at a higher speed than the rotor or low-speed shaft.

- Generator – Actual mechanism that converts the rotational kinetic energy into electricity.

- Wind Vane – Detects direction of wind and adjusts the rotor and nacelle to compensate.

- Yaw Drive – Keeps the rotor and therefore the turbines facing the wind.

- Tower – Elevates the aforementioned components to an altitude that optimizes wind exposure.

There are two main types of wind turbines that can be seen in design and implementation in the wind energy industry today. The first and most common type is the horizontal axis wind turbine that relies on a horizontal shaft that runs perpendicular to the blades which spin vertically. These wind turbine systems can be seen in use in major wind farms as well as solo operations.

The second type which is less common among the wind energy industry is the vertical axis wind turbine. As one might be able to infer, the vertical axis turbine has a vertical shaft in which the blades or rotor are connected to and spin horizontally. There are many variations of the vertical axis wind turbine but the major benefit is that maintenance is easier because the gearbox and generator are more accessible.

- Horizontal Axis Wind Turbine – This is the standard type of wind turbine where the low-speed shaft that connects to the rotor is horizontal. There are various ways to construct this wind turbine but they all follow the same concept as outlined above. The rotor spins with the wind and the rotational kinetic energy is converted to electrical energy through a generator.

- Vertical Axis Wind Turbine – This type of wind turbine is less common but has an advantage in that the rotor does not need to face into the wind. The shaft connecting to the rotor is vertical and the gearbox and generator are generally at the bottom of the tower. There are many types of vertical axis wind turbines all of which follow the same concept of force along the X-axis (parallel to the ground) as opposed to horizontal axis turbines which use force along the Y-axis (perpendicular to the ground).

The following are different variations that come from vertical axis wind turbine systems. Many of these were engineered decades ago and are no longer seen in use today, however the designs for these have been adapted and tweaked such that newer models can be developed that are more efficient with less problems than the older ones.

- Darrieus Wind Turbine – This vertical axis wind turbine uses curved blades that rotate and creates an internal force of wind that enables the rotor to spin at high speeds regardless of the wind speed. The downside of this is that this turbine generally requires an external motor to start spinning.

- Giromill – A variation of the Darrieus Wind Turbine in that it uses an H shaped rotor. The difference between the two is that the giromill uses straight blades that run parallel to the shaft. Other than that, the two operate on the entirely same principal.

- Cycloturbine – A type of giromill that not only has straight blades running vertically but also that the straight blade can itself rotate around its center axis. The advantage in this type of turbine is that it generates the most amount of power and can self-start (start without any external assistance).

- Savonius Wind Turbine – This vertical axis wind turbine relies on the principles of drag and wind resistance to function. The blades are shaped like an S with the two curved parts of the S moving with the wind. The curved part creates less drag and therefore the rotor is able to spin. These turbines do not generate much energy.

- Vortexis Wind Turbine – This is the most recent development of vertical axis wind turbines. It has seen use in Afghanistan and Iraq by special forces needing to power their devices. This turbine has two sets of blades, one smaller set that sits in a circle and one bigger set that surrounds the smaller set in a larger circle, that act in a gearbox like fashion. The outer set of blades use the wind to spin and by that set of blades spinning they force their own wind to turn the smaller inside set of blades. These blades are connected to the shaft which then turns a generator.

## Working Principle of Wind Turbine

The majority of wind turbines consist of three blades mounted to a tower made from tubular steel. There are less common varieties with two blades, or with concrete or steel lattice towers. At 100 feet or more above the ground, the tower allows the turbine to take advantage of faster wind speeds found at higher altitudes.

Turbines catch the wind's energy with their propeller-like blades, which act much like an airplane wing. When the wind blows, a pocket of low-pressure air forms on one side of the blade. The low-pressure air pocket then pulls the blade toward it, causing the rotor to turn. This is called lift. The force of the lift is much stronger than the wind's force against the front side of the blade, which is called drag. The combination of lift and drag causes the rotor to spin like a propeller.

A series of gears increase the rotation of the rotor from about 18 revolutions a minute to roughly 1,800 revolutions per minute - a speed that allows the turbine's generator to produce AC electricity.

A streamlined enclosure called a nacelle houses key turbine components - usually including the gears, rotor and generator - are found within a housing called the nacelle. Sitting atop the turbine tower, some nacelles are large enough for a helicopter to land on.

Another key component is the turbine's controller, that keeps the rotor speeds from exceeding 55 mph to avoid damage by high winds. An anemometer continuously measures wind speed and transmits the data to the controller. A brake, also housed in the nacelle, stops the rotor mechanically, electrically or hydraulically in emergencies.

### Wind Turbine Applications

Wind Turbines are used in a variety of applications – from harnessing offshore wind resources to generating electricity for a single home:

- Large wind turbines, most often used by utilities to provide power to a grid, range from 100 kilowatts to several megawatts. These utility-scale turbines are often grouped together in wind farms to produce large amounts of electricity. Wind farms can consist of a few or hundreds of turbines, providing enough power for tens of thousands of homes.

- Small wind turbines, up to 100 kilowatts, are typically close to where the generated electricity will be used, for example, near homes, telecommunications dishes or water pumping stations. Small turbines are sometimes connected to diesel generators, batteries and photovoltaic systems. These systems are called hybrid wind systems and are typically used in remote, off-grid locations, where a connection to the utility grid is not available.

- Offshore wind turbines are used in many countries to harness the energy of strong, consistent winds found off of coastlines. The technical resource potential of the winds above U.S. coastal waters is enough to provide more than 4,000 gigawatts of electricity, or approximately four times the generating capacity of the current U.S. electric power system. Although not all of these resources will be developed, this represents a major opportunity to provide power to highly populated coastal cities. To take advantage of America's vast offshore wind resources, the Department is investing in three offshore wind demonstration projects designed to deploy offshore wind systems in federal and state waters by 2017.

## Small Wind Turbines

Small wind turbines (SWTs) are a distinct and separate group of devices developed within the wind energy sector. According to the IEC 61400-2 standard, SWTs are characterized by a rotor area of <200 m2 and rated power below 50 kW. Wind power plants in this category are generally designed for small and individual customers such as households, farms, weather stations, road signalization, and advertising systems. SWTs offer a promising alternative for many remote electrical uses where, given a set of site evaluation criteria, the wind resources can be identified as beneficial, both as stand-alone applications and in combination with other energy conversion technologies such as photovoltaic, small hydro or diesel engines.

The quantity of SWTs operating worldwide grows every year. In 2012, the total number of such devices was approximately 800,000 worldwide with the growth of about 10%. The majority of SWTs (about 70%) are located in China, where the highest number of new installed units in 2012 was also noted. The second biggest market of SWTs is USA, where around 155,000 SWTs are operating at the time this document is prepared. In Europe, the leader is the United Kingdom: 23,500 units, followed by Germany: 10,000 units, Spain: 7020 units, and Poland: 3200 units. Total SWT generation capacity installed in 2012 was equal to around 678 MW (576 MW in 2011). The majority of world's capacity (85%) belongs to three countries: China (274 MW), USA (216 MW), and UK (83.7 MW). Unfortunately, developing countries play a minor role in small wind turbine industry. Electrical capacity growth in 2013 was small, with just 90 MW installed across Africa, for a cumulative total of 1255 MW. It is exceptionally regrettable considering enormous wind power potential (best around the coasts and in the eastern highlands of the African continent). A global forecast concerning SWTs installed capacity in years 2009–2020 is presented in figure.

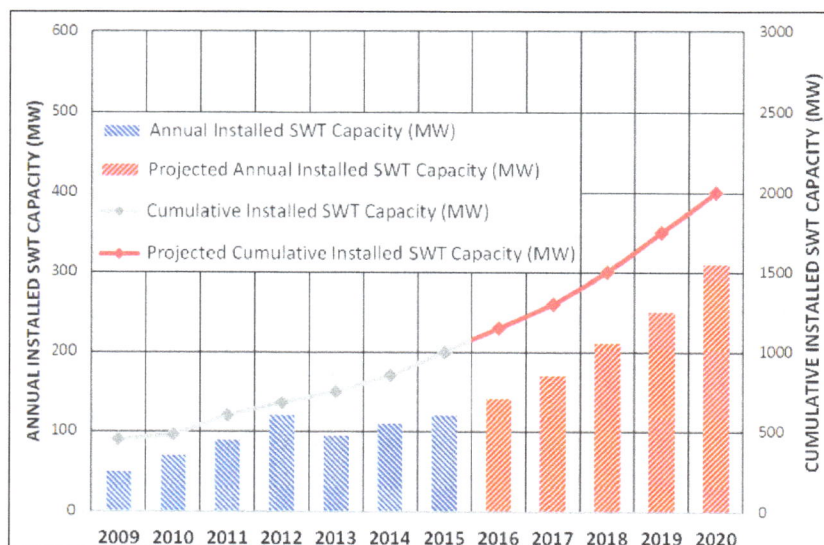

SWT installed capacity world market forecast.

The development and dissemination of SWTs involves great expectations in the field of eco- energy production. Some opinions suggest that without the dissemination of SWTs, the fulfillment of legal requirements for energy efficiency and energy production from renewable sources will be relatively difficult. In particular, in the developing countries, small wind turbine sector can efficiently contribute to provide electricity to millions of people in rural areas. In order to create a positive outcome, a big challenge awaits not only the authors of laws supporting investments in SWTs, but also engineers and scientists who should propose design solutions addressing the real issues found in the small wind turbine energy sector (aerodynamics: ill-optimized blade often using trends and observations from large HAWTs, poorly addressed issue of local wind resources abundance; structural: different nature of loads experienced by SWTs, little care given to weight optimization; conversion/control: seldom usage of active regulation methods, such as rpm control to expand operational margin of the SWTs, little attention given to match turbine's mechanical power capability with that of a generator; economic: lack of cost optimized SWTs).

The wide prevalence of SWTs and the emergence of the so-called "prosumers" within the electrical grid (the Smart Grid concept) are believed to be one of the biggest factors changing the way that the power companies will deliver service over the next decade. This qualitative change may result in a reduction of transmission losses and make the electricity infrastructure more flexible and secure. The specialists agree that work on the design and architecture of the future grid is as important as the work on the technologies and products that would realize a smart grid vision, for example, SWTs being a part of electricity network. The above motivates to devote more effort to the field of modern SWTs development.

In the small wind turbine market, the return on investment (ROI) is one of the most important factors determining the turbine's validity. Having the above in mind, it has become a challenge for many designers and research facilities to develop a small wind turbine design which would be competitive with other renewable energy sources. For this to be possible, it would have to incorporate a number of factors: high efficiency, sufficient longevity, low installation, and maintenance costs. Having above in mind, it has been concluded that SWTs should be characterized by the lowest possible price, while maintaining relatively high efficiency as well as satisfying reliability and maintenance parameters.

## SWT Design Solutions

This topic presents both the literature investigation including manufactures data and independent expertise as well as series of coupled numerical simulations (computational fluid dynamics (CFD), Finite Element Method (FEM), multi-body), investigating impact of the chosen design solutions on small wind turbine operation. It allowed objective evaluation of different design approaches (advantages and disadvantages), which in turn enabled the systematic identification of actual limitations as well as the opportunities for specific design solutions of SWTs: horizontal axis wind turbines (HAWTs) and vertical axis wind turbines (VAWTs); the rotor position in relation to the tower (upwind vs. downwind); and the addition of a duct that encapsulates the rotor (diffusor-augmented wind turbine—DAWT).

## Horizontal and Vertical Axis SWTs

Wind turbines in general can be divided into two major groups based on the position of rotor axis of revolution: HAWTs and VAWTs.

HAWTs are currently the vast majority of installed solutions worldwide. Consequently, close to 75% of SWTs are HAWTs. That is due to their high efficiency in relation to low installation and maintenance costs. Apart from the unconventional solutions and one- and two-bladed configurations, the classic three-bladed layout is the default choice with a 99% market share of all operating HAWTs. Three blades is the lowest blade count which allows for easier management of centrifugal body forces. Some small turbines incorporate an increased amount of blades, which may improves aerodynamic efficiency slightly, but most often the added mass and material cost outweigh any benefits gained.

VAWTs are the earliest examples of harnessing the energy of the wind. These early machines were primarily based on a drag force as the mechanics to turn the turbine's working section. Some modern designs try to re-invent the concept, but the efficiency is generally unsatisfactory because of the limited tip speed ratio. Modern-day VAWTs most often use the aerodynamic principles of lift force to work, with Darrieus wind turbines being a good example. Some of the marketed advantages of VAWTs over the classic horizontal designs are as follows: easier maintenance thanks to the generator and the gearbox being placed near the ground, low cut- in wind speeds and no need for yaw mechanisms. In practice, maintenance and operational costs are similar because of the high loads on the bottom bearing, efficiency in converting wind kinetic energy into a mechanical power is relatively low for low wind speeds and response to an often changing wind direction dominated by turbulence is poor. Nevertheless, VAWTs are a popular choice for urban wind environments and are seeing some optimization research with advanced evolutionary algorithms aimed at raising the overall performance of these designs.

## Shrouded SWTs

An interesting design solution for HAWT turbines is the addition of a circular duct that encapsulates the rotor. Such turbines have a number of commercial and scientific designations, such as DAWT, wind lens, compact wind acceleration turbine (CWAT). Attempts to add a shroud in order to stabilize and accelerate the flow through a turbine have been reported nearly 150 years ago. An invention proposed by Ernest Bollée, patented in 1868, was an American style multiblade wind turbine with a stator in the form of a shroud. After decades of stagnancy, shrouded turbines have seen a major interest increase in recent years with many academic and industrial centers proposing mechanically and aerodynamically optimized solutions which are said to be vastly superior to non-shrouded designs. The influence of the shroud on the air flow has been proven to be beneficial in numerous wind tunnel tests, but little is available on the performance of such turbines in real outside conditions. What is more, the added diffusor is often expensive due to the amount of material used and the added mass puts more stress on the towers foundation and hinders the operation of the yaw system.

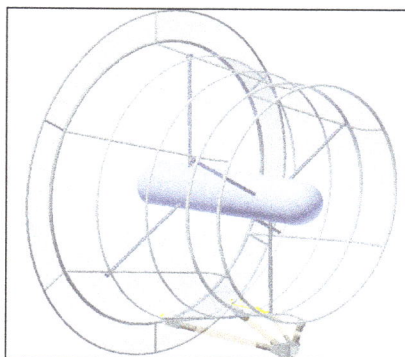

Framed light shell diffuser concept for SWT.

The concept of framed light shell diffuser was created as an attempt to lighten the diffuser made of glass fiber composites which are the most commonly used materials for such applications nowadays. Geometrical shape of the diffuser was developed based on CFD simulations considering turbine efficiency optimization for low wind speed conditions. The authors focused on developing an overall structural design, as well as detailed technical solutions (stiffer struts, connecting rods, material forming issues). Subsequently, static structural finite element analysis was performed in order to assess the stiffness and stress distribution based on load conditions in the form of pressure field form the CFD analysis and gravitational forces. Obtained results lead to a conclusion that designed aluminum shell frame diffuser may be an alternative for the composite diffuser. Moreover, the developed structure is characterized by lower mass and comparable stiffness to its composite counterpart. It is worth mentioning that lighter diffusers allow use of lighter supporting structures, such as towers or guy-wired masts. This fact may contribute to lowering the overall costs of future SWTs and could be very beneficial for the privately owned small wind turbine market.

## Upwind vs. Downwind SWTs

One of the fundamental distinguishing factors in horizontal wind turbines is the rotor position in relation to the tower. Today market is dominated by upwind rotors: relative to the wind, their rotor is located in front of the tower (windward), while downwind rotors are behind the tower (leeward)

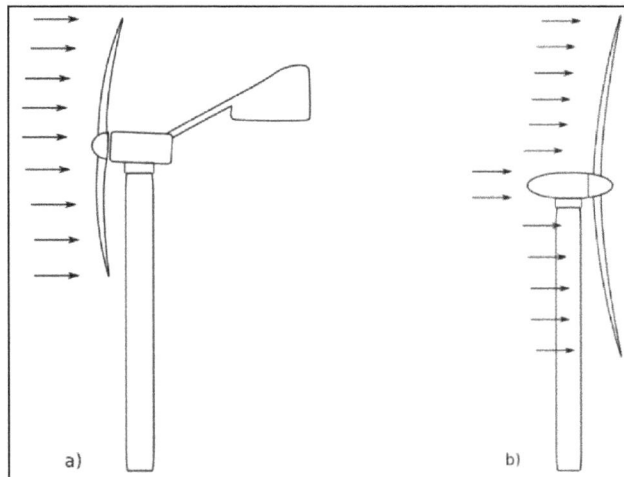

Upwind (a) and downwind (b) wind turbines.

In upwind design solution, the rotor is in the front of the unit (facing the wind) and it is characterized by higher efficiency due to the reduced tower impact on wind inducted at the working section. Unlike upwind, the downwind turbines experience a different inlet wind profile effectively changed by the development of a boundary layer on a nacelle and by the formation of wakes in the aerodynamic trail behind the mast. In upwind designs, the rotor needs to be made rather inflexible and placed at some distance from the tower to avoid collision. What is more, those are not self-aligning in the direction of the wind, and therefore, they need a tail vane or a yaw system to keep the rotor facing the wind.

In downwind turbines, the rotor is on the back side of the turbine (the lee side of the tower). Their main advantages are that they may be built without a yaw mechanism and the rotor may be made more flexible (since there is no danger of a tower strike). This may be an advantage both in regards to weight and the structural dynamics of the machine. Therefore, the basic advantage of the downwind machine is that theoretically, it may be built simpler and lighter than an upwind machine. Downwind turbines generally have lower aerodynamic efficiency, and the fluctuation in the wind power due to the rotor passing through the wind shade of the tower may give more fatigue loads on the turbine than those with an upwind design. Although the upwind type is more popular than the other one, the advantage that the downwind configuration can face the wind automatically makes them much more promising for SWTs due to its simplicity.

Using the methodology of aero-, servo-, elastic-coupled numerical analyses, the authors carried out analysis of small wind turbine behavior in terms of upwind vs. downwind SWT comparison. It is worth pointing out that the relative low cost of such research methods is a great advantage in comparison to real experiments (e.g., wind tunnel tests).

The first set of the obtained results concerns the dynamic response of the simulated variants in terms of rapidly changing wind speed conditions. Case one was here the IEC 61400–1 direction-change condition. Both considered turbine variants have proved to be capable of quick and precise nacelle turn during the changing direction, event and it can be stated that both design solutions work properly and similarly in this case. The upwind variant of the small wind turbine proves to have 5% higher efficiency than the downwind variant, which in turn resulted in a 13% higher energy output in economic analysis. For the 3 kW machine, the downwind variant should be at last 1200 USD cheaper than the upwind small wind turbine to be economically valid (by the application of lighter and cheaper materials, deflective blades, etc.); on the contrary, it is more likely to equip the small wind turbine in a tail vain having additional 1200 USD. The biggest disadvantage of the downwind design variant is the presence of significant fluctuations of momentums in rotor blades and forces on the top of turbine tower, which may cause a real danger of fatigue damage in the turbine construction as well as the risk of resonance.

## Other SWTs Designs

Other design solutions that can be found in SWT applications or concepts are, that is, as follows: two-blade, single-blade turbines with a counterweigh, multiblade rotors, Venturi design turbines, Magnus effect turbines, multi-rotor solutions, H-type turbines, Darrieus–Savonius hybrid, Tornado/Jet/Vortex/Spiral airfoil, Fuller's design (bladeless—Tesla like), Pawlak's design, airborne concepts, and other unconventional solutions.

The main driver for multiblade (up to three) turbine development is the fact that aerodynamic efficiency increases with the number of blades. Increasing the number of blades from one to two yields a six percent increase in aerodynamic efficiency, whereas increasing the blade count from two to three yields only an additional three percent in efficiency.

Extraction of wind energy by a single rotor leaves a substantial amount of power unrecovered. To use this remaining potential, a two-stage wind turbine was proposed. Contra-rotating wind turbines possessing two co-axial rotors can theoretically gain up to 40% more energy from a given swept area as compared to a single-rotor turbine. Contra-rotating turbines require a generator tailored for this system in order to avoid expensive and impractical placement of the second generator on rotor-nacelle assembly (RNA) to convert energy from the additional rotor. In fact, the twin shaft technology of co-axial rotors presents a possibility to increase the rotation speed of the electrical generator by summing up the relative velocities of the rotor and stator. The main drawbacks of two-stage turbines are an increased interaction between the rotors posing problems from aero-mechanical point of view and the additional costs associated with the installation of the second stage and a more sophisticated generator.

The Savonius rotor is a self-starting, high-torque wind turbine. It may be used alone or to jump start the Darrieus rotor, a high-efficiency rotor, but with a limited capability to initiate operation on its own. This combination is presented as an effective design that combines the advantages of both designs.

The Magnus effect is the commonly observed effect in which a spinning ball (or cylinder) curves away from its principal flight path (a force perpendicular to the direction of movement, acting on the rotating cylinder or other rotary body, moving relative to the fluid). This makes a range of potential advantages with respect to traditional blade wind turbine. Radial cylinder location is analogous to wind wheel blades with horizontal axe. The basic advantages are said to be seen at low, but the most repeated wind velocities 2–6 m/s, at which blade wind turbines are not effective.

## SWT's Blades

The blades are the components, which interact with wind and are designed to maximize the turbine efficiency. Blades are made from light materials, such as glass- or aluminum-based fiber-reinforced plastics, possessing good resistance to wear and tear. The fibers are incorporated in a matrix of polyester, epoxy resin, or vinyl ester constituting two shells kept together and strengthened by an internal matrix. The external surface of the blade is covered with a layer of colored gel to prevent ageing of composite material due to ultraviolet radiation.

A hollow shell corresponding to the defined blade envelope clearly provides a simple, efficient structure to resist flexural and torsional loads, and some blade manufacturers adopt this form of construction. The hollow shell structure defined by the airfoil section is not very efficient in resisting out-of-plane shear loads, so these are catered for by the inclusion of one or more shear webs oriented perpendicular to the blade chord.

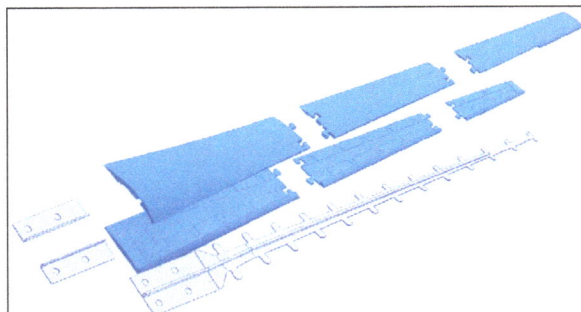

3D printed ABS polymer-based SWT blade with a stiffening core made of steel.

Five different materials were considered in terms of turbine blade optimization: steel, aluminum alloy, glass fiber composite, carbon fiber composite, and ABS polymer for three-dimensional (3D) printing. Blades made of all listed materials have fulfilled the strength criterion (stresses values generated by maximum loads did not crossed yield or fracture). Analysis revealed that the best material in terms of mass and stiffness is the carbon composite. Unfortunately, this material is very expensive and was opted with in order to keep SWT price reasonably low. Metal-based materials are characterized by very high stiffness, which reduces the blade tip deflection; however, the mass of the blade was too high in this case. What is more, a relatively high stiffness was a direct cause of relatively high stress values. Those values have raised justified questions on the fatigue toughness of the SWT. The ABS polymer is the cheapest and the lightest one of all mentioned. It is also the best choice from the manufacturing point of view. Unfortunately, it is also the most flexible material, producing high blade tip deflections even after the introduction of stiffening core made of steel.

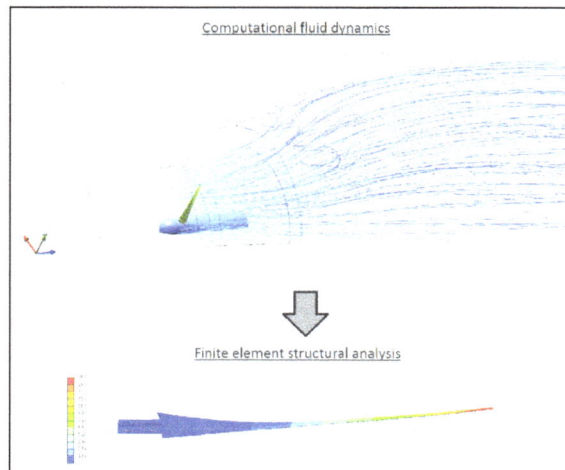

SWT blade deflection (ABS 3D printed coating with a steel core).

## SWT's Generators

SWTs are usually equipped with synchronous generators operating at variable angular speeds based on permanent magnets (e.g., neodymium magnets). It is a design solution which can contribute to meeting the requirements of efficient operation. Working principles of the variable speed generator determine many other aspects of the wind turbine operation including, for example, a tremendously increased annual energy production (AEP) in comparison with equal-power fixed pitch, fixed speed turbines.

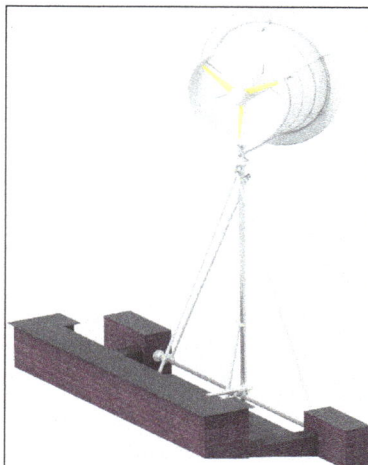

Roof-mounted SWT equipped with the diffuser.

SWTs typically operate in autonomous systems without connection to the grid, in which they directly power heating and other loads that do not require stable electric parameters, dictated by economic issues. Potential expenditure related to connecting the SWT to the electrical grid exceeds the usual financial benefits of possible income from selling energy produced. The other aspect is the fact that SWTs are generally simple devices and they are rarely equipped with sophisticated control systems such as blade pitch mechanisms.

This type of generator produces alternating current (AC), which must be rectified to direct current (DC) by means of a simple bridge rectifier. Similar to solar photovoltaic (PV) systems, the DC voltage allows the use of these turbines for battery charging. For in-battery charging systems, a charge controller is added to prevent the battery from overcharging. A dump load is required to protect the inverter from overvoltage and to prevent the turbine from over-speeding.

## SWT's Supporting Structures

The SWT is usually placed on a pole, preferably higher than 15 m, to lessen the impact of turbulent, sheared winds forming in close proximity to the ground. Utility-scale wind turbine towers are mostly of tubular type. In case of SWTs, tubular type towers are also popular; however, guyed lattice towers are also used. The tilt-up poles/towers are very popular since they are easy to install and offer good accessibility for maintenance and repair. Tubular towers are usually made of rolled steel, although sometimes reinforced concrete is used.

Rooftop-mounted turbines are yet another option. Such turbines are directly installed on a building. Both vertical and horizontal axis turbines can be mounted in such a way, but they might be subject to slightly different wind profiles being inducted at the rotor. Particularly important for roof-mounted turbines is the identification of flow separations due to rooftop edges. Such flow structures not only influence the production efficiency but may be a source of unwanted loads that are difficult to control.

## SWT's Economic Evaluation

The investment in SWTs is more often made by private individuals in order to partially fulfill residential household demand for electricity or to produce hot tap water. However, practice shows that a large portion of those who have decided to build a backyard wind turbine is disappointed with the actual amount of energy produced by it. This is mainly due to the promotion of inaccurate, overstated, or even incorrect information on the projected turbine's power output by manufacturers or installers. This reduces public confidence in the legitimacy of this sort of economic investment, thus slowing down the development of this industry sector.

Numerical weather prediction (NWP) simulations over the span of year 2013, covering the whole area of Poland, were carried out. The purpose of these forecasts was to establish a reliable and accurate wind resource atlas for approximate AEP of small wind turbine systems. Long- term wind speed forecasting has been used with success for utility-scaled wind farms. This accounts not only for planned sites but also operational power plants, where weather fore- casting is used constantly for setting up advanced control schemes or predicting suitable time windows for planned maintenance.

Calculation of meteorological parameters was performed using the non-hydrostatic mesoscale Weather Research & Forecasting (WRF) model. WRF is designed both for operational fore- casting and atmospheric research use. It enables an atmospheric phenomena simulation for scales ranging from thousands of kilometers to single meters. The model contains several capabilities, important from the research point of view, such as: 3D data control with initial data and model results correction possibilities using aerological, radar and satellite measurements; multilayer ground modeling; humidity cycle; vegetation; cloud cover and precipitation parameterization including water phase state; radiation transport processes; and boundary layer with turbulent vertical transport. These modules are responsible for the parameterization of physical phenomena occurring in the atmosphere. The boundary layer area of intensive surface radiative forcing and mechanical forcing is an important part of the simulated phenomena. Mechanical forcing is determined by orography, roughness, and cover. Radiative forcing is defined by albedo and thermal capacity of the surface. The parameters are deter- mined by the ratio of land to water area; vegetation and its status; irradiation depending on location and inclination. Determination of these factors was of importance because the boundary layer close to the surface is the area associated with the presented studies.

The data used in the study covered the period of 1 year. The input data of the WRF came from the archive of the global forecast system (GFS). These were obtained from the National SOO Science and Training Resource Centre (NWS SOO/STRC). The model was run for each day and for each main synoptic term. Two grids with spatial resolution of 36 and 12 km were used. The simulation domain included a selected area of Europe as shown in figure. Forecast modeling time was set equal to 24 h with a 1 h data sampling rate. Results for each 24-h period were obtained in a five-dimensional set of prognostic parameters fields. This set includes, among others: pressure values, geopotential, tem- perature, and three- dimensional wind fields. Wind speed data averaged over a 60 min period have been said to be useful for long-term energy yield predictions for SWTs.

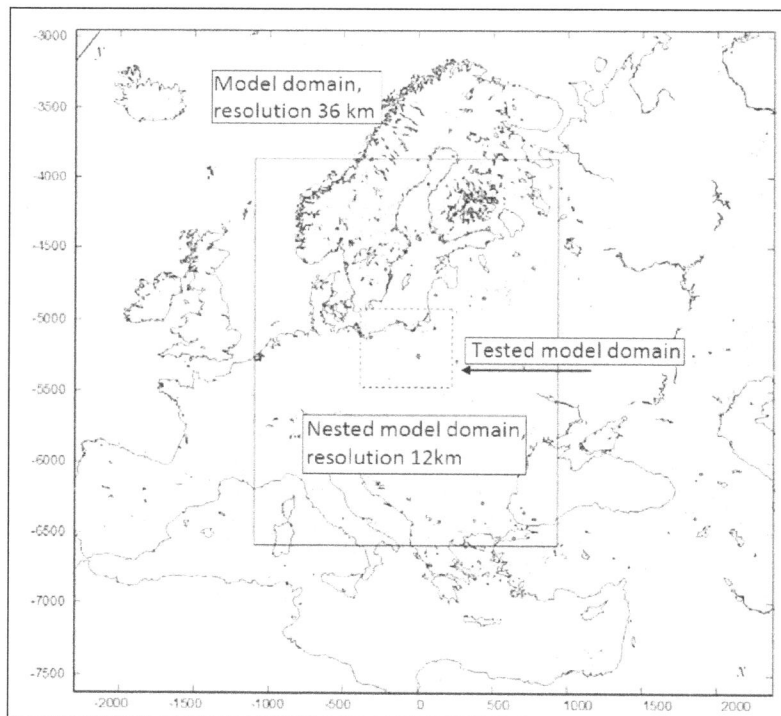

WRF model domains—36 and 12 km grid.

A randomly chosen location in model domain was a rural region neighboring a small town called Sokolow Malopolski in the Voivodship of Sub-Carpathia. It is characterized with rather typical rural wind speed conditions. Many of the houses built in the region are solitary and surrounded by open grass or farm fields, thus producing a suspected low ground roughness rate. Such households could potentially benefit from an auxiliary power or heating source in the form of a small wind turbine.

Calculations of the power output were performed using the following formula:

$$E = \sum_{i=1}^{n} \begin{cases} \frac{1}{2} A \rho_i \eta(v_i) v_i^3 h & P_r > \frac{1}{2} A \rho_i \eta(v_i) v_i^3 \\ \\ P_r h & P_r < \frac{1}{2} A \rho_i \eta(v_i) v_i^3 \end{cases}$$

where: E is the estimated total energy generated throughout the test period, n is the number of samples, h is the sample duration (1 h), vi is the wind speed for sample i, A is the swept area of a given turbine, $\rho_i$ is the air density for sample i, $\eta(v_i)$ wind turbine efficiency at given tip- speed ratio (TSR) and $P_r$ is the turbine rated power.

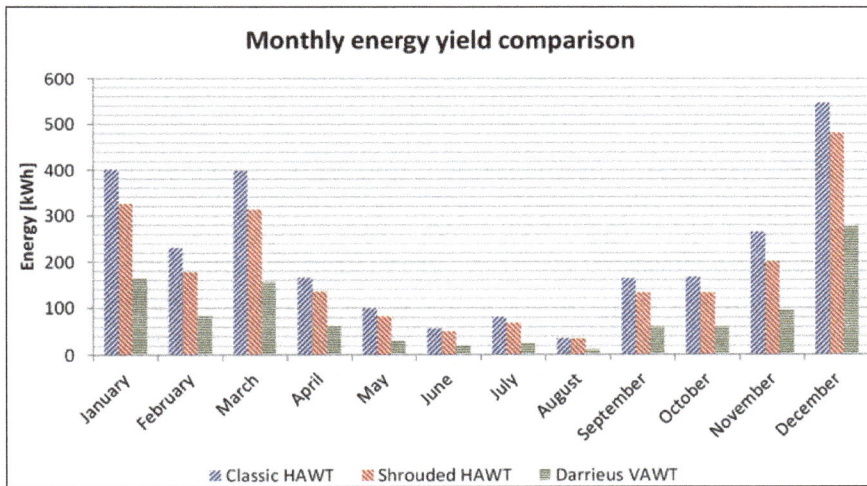

Monthly energy output for the chosen turbines.

## HAWT, VAWT and DAWT Comparison

Three types of SWTs were chosen to be compared in terms of their productivity and cost- effective- ness. The first is a typical three-bladed HAWT design. The second is an aerodynamically complex, shrouded HAWT. The third is the Darrieus-type VAWT. All the turbine characteristics, such as power curves, price, swept area, were taken from very popular, commercially proven designs. For the results to be meaningful, all the chosen turbines have a similar rated power of around 3 kW.

The outcome of the performed calculations is shown in figure and compared in table. When esti- mating the investment return period, the price for energy supplied (power supply) was established at 0.2 \$/kWh. Maintenance and any additional costs associated with turbine operation are as- sumed to be the same for all three systems and are therefore neglected.

Table: Chosen designs economic comparison.

|  | Energy produced (kWh/year) | ROI (years) |
|---|---|---|
| Classic HAWT | 2610 | 28+ |
| Shrouded HAWT | 2126 | 31+ |
| Darrieus VAWT | 1036 | 137+ |

The difference in monthly power generation is clearly visible with the majority of energy being produced during the windy winter months. Any needed maintenance or routine inspections should be carried out during calm summer months. Such a discrepancy in electricity generation raises the problem of continuous power supply.

For the chosen localization, the classic three-bladed HAWT proved to be the most cost-effective design of those taken into account, even though it had the lowest rated power. In theory, the advanced construction of the shrouded HAWT turbine allowed for a very high theoretical efficiency. In practice, because of its small rotor diameter it proved to be slightly inferior to the classic layout with a larger rotor. Moreover, the shroud is likely to put additional strain on the tower due to added mass and increased aerodynamic drag, possibly increasing the maintenance costs. Snowfall and icing of the shroud could lower the operational availability and introduce the need for deicing. The chosen Darrieus vertical wind turbine was the worst design in the terms of cost-effectiveness. Very high system costs and low efficiency in relation to rotor size led to the return period being unacceptably long.

## Investigation of Parameters Influencing the Efficiency of SWTs

Three cases will be carried out in order to find the factors which have the highest impact on the estimated electrical power generation. Case one investigates the difference between estimated one with a theoretical, constant efficiency turbine and a popular, commercially available, wind turbine using its power curve. The second case estimates the generated amount of energy based on three types of wind data for a randomly chosen site. In this case, calculations were performed using a yearly mean wind speed, discretized meteorological data and the Weibull wind speed distribution function. The third case shows the difference of estimated energy production for three tower heights: 10, 20, and 30 m. Calculations of the power outputs were performed using formula.

The turbine chosen for the analyses is the Evance R9000, which is a popular small wind turbine design with a high market share, especially in the UK. It is a 5 kW turbine, with a 5.5-m-radius rotor and a typical, of HAWTs, power curve. Factors such as maintenance and down time were not taken into account. That is due to the fact that the goal for these calculations was to estimate small wind turbine sensitivity to operating parameters such as a tower height, efficiency, and wind data type with which the AEP forecast is modeled.

The estimated energy produced by the Evance R9000 amounted to 7551 kWh, assuming energy cost of 0.2 €/kWh, that is an equivalent of 1510 €. For a 5-year ROI period, without any maintenance or down time, the turbine would have to be priced at around 7500 €; however, in reality, such solutions are roughly priced at 36,500 €. Another aspect worth mentioning is that a yearly demand for energy of typical household is estimated at around 3–4 MWh. Thus, in theory, it would be possible to achieve energy independence. In practice, it would be nearly impossible.

Another very interesting way to present the obtained results is to draw an energy density function on top of the occurring wind speeds. Figure shows, in blue, the wind speed occurrence rates and, in red, the amount of energy generated from a specific wind speed by the Evance R9000 turbine during the whole year. The shift of the energy density function to the right in respect to wind speed curve is a direct result of the power equation, proportional wind speed cubed equation:

$$E = \sum_{i=1}^{n} \begin{cases} \dfrac{1}{2} A \rho_i \eta(v_i) v_i^3 h & P_r > \dfrac{1}{2} A \rho_i \eta(v_i) v_i^3 \\[2ex] P_r h & P_r < \dfrac{1}{2} A \rho_i \eta(v_i) v_i^3 \end{cases}$$

Using Matlab computer software, the maximum likelihood estimate of the Weibull parameters has been established with the wlbfit function. The resultant Weibull probability density function is shown in figure. Additionally, an arithmetic mean of the wind speed for the given location has been calculated and is equal to around 4.77 m/s. The yearly power generated has been again calculated using the obtained Weibull function and the mean wind speed with the standard Evance9000 power curve. A comparison of the estimated power production based on three different wind data types is presented in figure. It is clearly visible that estimating the amount of wind power generation using an arithmetic mean of the wind speed is highly inaccurate. The predicted amount was 4420 kWh which is almost 70% off the amount of energy predicted using the direct method with accurate wind speed data. Energy production estimated with the Weibull function equaled 7932 kWh, which is 5% different from the result obtained using the direct method. This difference is small enough to prove the validity of using the Weibull or Rayleigh distributions for estimating small wind turbine energy production.

Five different calculations of estimated energy output were performed with five different small wind turbine characteristics. The first was performed with an artificial, constant value of 33%. Three characteristics were based on the Evance R9000 power curve including the original curve, one which is offset by +5% and another which is offset by −5% from the original distribution. The fifth one is a hypothetically proposed, small wind turbine curve which has increased efficiency in low wind speeds—STOW. The results of the performed calculations are presented in figure.

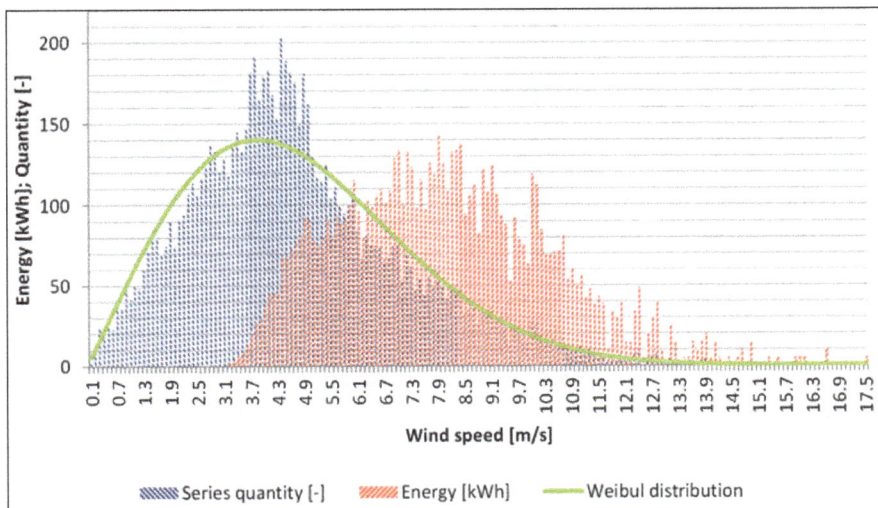

Wind speed distribution, energy distribution, and Weibull function fit (c = 5379; k = 2023).

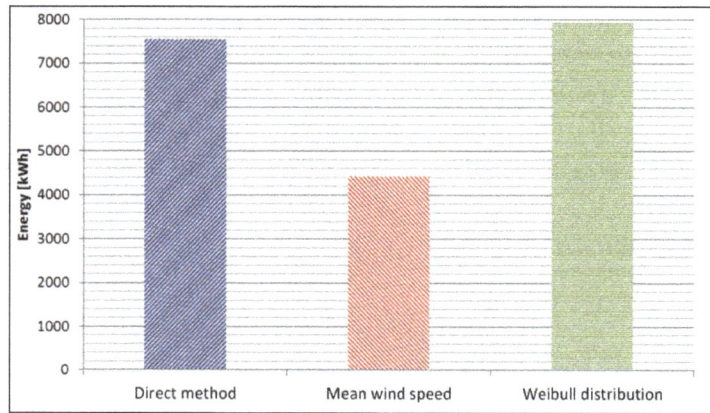

Energy production estimates based on three different wind speed data types.

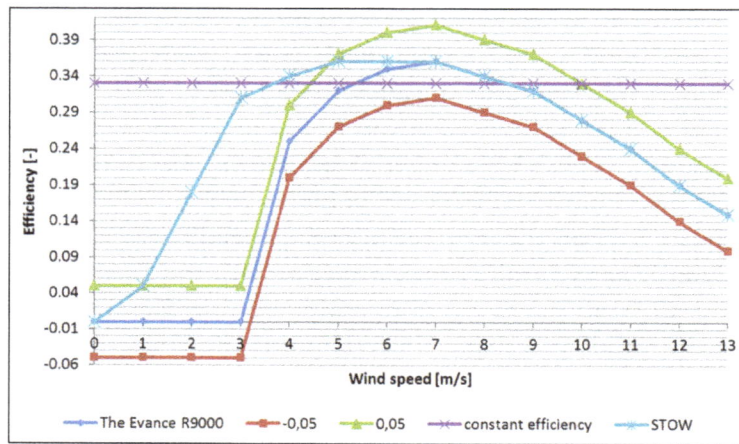

Efficiency curves used for the energy calculations.

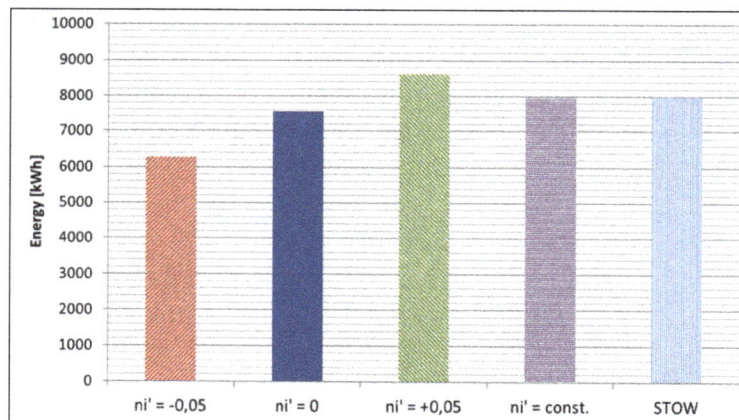

Energy produced by a 5 kW rated turbine with different
efficiency curves; the series are tied to the curves presented in figure.

The final investigated small wind turbine parameter was the tower height. Wind speed data were collected at 10 m height. Using the power law with the terrain dependent parameter $\alpha = 1/7$, the wind speeds for the whole year have been recalculated for 20 and 30 m heights. The obtained wind speeds were used to estimate the produced power with the standard power curve and are presented in figure.

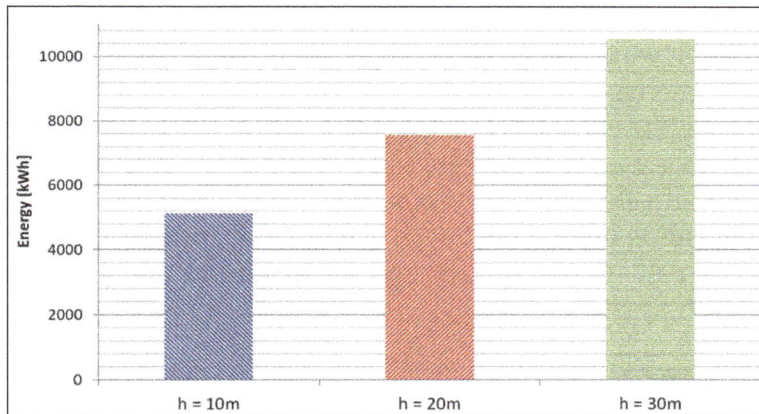

Estimated energy generated for different tower height variants
(10, 20, and 30 m)—wind speed obtained with the empirical power law.

# Applications of Small Wind Turbines

Wind energy is considered a very promising renewable energy source, estimating to cover the 20% of the global electrical energy demand in 2020, while several technical improvements have been investigated and new fields of application have been opened. Apart of large size WTs, which are widely applied being directly connected with the electricity grid, small size WTs have been improved. These wind generators can be effectively applied at locations of sufficient wind energy potential, mainly in rural and remote areas, being also suitable for agricultural and urban applications. In several applications, the WTs can be combined with Diesel Generators to cover the electricity needs in case of low or no wind. Solar energy and wind energy are renewable energy sources that can be effectively combined to produce electricity by photovoltaics (PV) and wind turbines (WT) and also heat by solar thermal collectors (T). Photovoltaics and thermal systems can cover hot water, space heating – cooling and electricity needs in many applications. The effective use of these systems is based on the hybrid PV/WT concept where PV and WT subsystems supplement each other in electricity production and that the electricity of both can be conversed into heat to cover the thermal load.

The hybrid photovoltaic/thermal (PV/T) systems are interesting alternative combinations and as solar and wind energies vary with time, energy storage (batteries, compressed air or hydrogen in the future) could be used to adapt the time of energy conversion with the demand profile. In these combinations, the WTs can have effective operation in windy days and nights and the solar energy systems in sunshine days. In addition, the surplus electrical energy from WTs and PVs can be converted into heat, if the grid cannot accept the produced electricity and the batteries are fully charged. Therefore the effective use of the three (WT, PV, T) subsystems can cover a great part of energy load, contributing to the conventional energy saving and environment protection. We also note that in off-grid or mini-grid applications, thermal needs as water heating is covered by electricity and the cost of it from WT or PV systems is competitive to that of Diesel Generators.

Last decade, among the publications on wind energy, many interesting works on the characteristics and the performance of small wind turbines are reported and this is the result of the augmented interest in this new area of wind energy technology, compared to medium and large wind power systems. The small WTs can be of horizontal or vertical rotor axis and their maximum power and dimensions must be taken into consideration in their applications. In horizontal axis WTs (HAWT) the tracking of the wind direction is necessary, which could be considered as a major disadvantage of this type WTs. Alternatively, stationary mounted HAWT systems have been suggested for several applications, including the modified type of wind concentrator. Vertical axis WTs (VAWT) are practical, as they use a fixed rotation axis and the turbine at the bottom, but they are still of higher cost compared to HAWTs.

An extended research of small wind turbines as long as development issues have been presented in a number of articles. The modified type of wind concentrator has been proposed and studied for the improvement of system output. The starting behavior of a small horizontal-axis wind turbine is a very important issue which may have a great effect on its performance and has been thoroughly examined by scientists. Vertical Axis WTs (VAWT) are more practical, because of the fact that they use a fixed rotation axis and also due to the low position of their motor- generator. However, they are still of higher cost compared to HAWTs. The Savonius and Darrieus Wind turbine types are the most usual VAWT systems, a combination of these two types has also been suggested and small commercial models are already out in the market.

During the last decade, more and more scientists and researchers have taken under consideration the prospects of using small wind turbines for energy supply in various applications. In other words, small wind turbines, may change radically the philosophy of modern world's energy system. The most important concern of small wind turbine industry is to construct systems that are cost effective and yet well designed so that they can withstand 20 or 30 years of operation in weather that is often severe. Because of the need for simplicity and high reliability, small wind turbines are still a difficult field for wind energy technology. Important issues such as the furling behavior, the thrust measurements, the yaw behavior, the blade and tower loads of small wind turbines remain subjects that need further research. Despite this, small wind turbines have already entered the commercial world and the practical aspects of their applications can only make optimistic, the people who range themselves with wind energy technology.

## Application of Small Wind Turbines

## General Aspects

To begin with, farms and district houses can become less or no dependent to the known and commonly used electricity grid. This could be done with the help of stand alone or grid connected small wind turbines that can provide energy for the everyday needs by these small units. Both wind turbines as stand alone systems and grid connected have been studied, while theoretical and experimental results have been presented. Small wind generators, ranging from 400 W to 40 kW or more, can cover the energy needs of an entire farm. In addition to this, grid connected WTs, enable farmers to get the maximum energy output of their wind generators. When the WT produces more power than the farm needs at the moment, the extra power can be supplied to the electricity grid. In this way, the farmer may not only cover his farm's energy needs but he can make a profit by selling power generated by the WTs installed in his farm. With smaller wind turbines, farms and houses use all of the produced energy. Nevertheless, the cost per unit of electricity generated from larger turbines is lower that that of the smaller ones. In this case, financial balancing to the expenses of wind turbine installation may take longer but, considering that a well maintained wind turbine can last 30 years, it surely can be very profitable for the investors.

## Telecommunication

Wind energy turbines can be also used for generating the power needed by small telecommunication stations, which are placed in remote and hardly accessible regions such as the small meteorological stations on the top of the mountains. These kinds of systems may be placed in areas where grid connection could be a totally unadvisable action. A small WT system could supply the systems with the necessary energy, while in case of having extra energy generated, it could be stored in batteries and be used when the meteorological conditions would forbid energy generation. A very important tool for the study of the wind characteristics of a region and consequently, of the performance of a wind turbine is Weibull model by which, a small number of days is used as a sample of a whole month's period and afterwards, using this model, the measures of these days can give us a very good view of a region's wind potential.

## Urban Applications

Small wind turbines, designed for urban use, offer today another very important prospect of wind energy applications. Small, building integrated wind turbines literally change the idea of energy production in the cities, make energy consumers less dependent on the grid and can be financially within reach for users. A lot of scientists study wind inside built areas and its characteristics, and try to find ways of exploiting wind fluctuation in urban regions. Phenomena, such as turbulence, play a significant role in the way the wind moves inside urban areas and that is why it is one of the main subjects of modern research. VAWTs seem to be the best choice for urban use as they are less sensitive to turbulence and wind changes than the HAWTs and also they have low weight relative to the power output. In addition to this, the fact that their motor is at low position, below the rotor, is a very important issue when we are interested in roof mounting and service access. Besides, VAWTs are not as visually attractive as HAWTs. Regardless to that, a key technological issue for the use of wind turbines installed on buildings is the type of the building and according to a research carried out by ECOFYS and Delft University of Technology, there are five building types, three of which are best suited for urban turbines: The wind catcher, the wind collector and the wind sharer. Observance of building regulations, design that suits the surrounding buildings, habitant's safety, wind turbine noise and the methods of installation and maintenance are still subjects for discuss and further research.

## Greenhouses

Greenhouses are another field where the use of small wind turbines could represent a great solution for their energy needs. Small wind turbines, installed nearby greenhouses, can be of horizontal or vertical rotor axis and must be of low cut-in wind speed. The output of the wind turbine depends on the wind speed at the location of the installation and it is obtained any time of the day or night that the wind speed is over a certain limit. A conventional electricity generator is usually necessary, in additional to the wind power system. The height of the turbine, its structure and color must be well designed in a way that maintains the harmony with the environment.

## Desalination Plants

Fresh water is extremely valuable in places like small, remote islands. Drinkable water can be either transported, with costs that are really high, or produced with the method of desalination. Desalination can be achieved with electrodialysis, vapor compression or reverse osmosis, which is the preferred technology because of the low energy consumption. Small wind turbines have been proposed and installed for the production of the necessary energy for desalination plants. In this way, fresh water transportation costs and fuel costs for the production of energy consumed by the desalination systems are avoided. The optimization of a wind turbine's size and performance in the combination with a desalination system are very important subjects for the cost-effectiveness of the application. Another important issue occurs when power variations, caused by the wind turbine performance, affect negatively the desalination system's lifetime. In these cases, back-up battery or Diesel generators are recommended.

## Water Pumps

A very important use of small wind energy turbines is already evident in many developing countries. In these countries, water pumping for drinking-water supply and farm irrigation is very popular and in many cases necessary. The design and application of a wind turbine that could be combined with water-pump systems can be a profitable and environment saving solution. Correct study of the location and water pump's features are needed for the construction of the wind turbine with the best possible performance and result. Regarding this, many different types of small WTs have been already combined with water pumps, offering a well functioning energy source.

## Hydrogen Production

New wind energy technology has make progress in the promising field of wind-hydrogen systems. Hydrogen is not an energy source as such, rather an energy carrier that enables its storage, transport and use. Used through fuel cell technology, hydrogen does not produce emissions and has no effect on global warming, as it only emits water vapor in the process of energy conversion. The electricity produced by a wind turbine, can be used for the generation of hydrogen and oxygen from water by an electrolyser. So, when excess wind energy is available, power produced is supplied to an electrolyzer which consumes water to generate hydrogen, which is afterwards stored as a compressed gas. Hydrogen can be also stored in other ways such as liquid or as a metal hydride. Compressed gas storage is currently the most cost-effective. Because of the fact that fuel cells (which are used for the conversion of hydrogen to electricity) are still a largely experimental technology, the costs for wind hydrogen are relatively high when compared with wind-battery and wind-diesel systems.

Nevertheless, hydrogen produced by the means of wind energy is a product with zero emissions throughout the production cycle of its own. It can be considered an ideal solution in the near future for the passage from fossil fuel generated energy to clean and renewable sources based energy, especially in remote applications. That is because remote locations have lack of utility infrastructure and high cost to extend or upgrade the electric grid for basic electric service, high cost of delivered fuel for conventional fossil fuel electric generation and low level of reliability of the current electric generation system. When there is sufficient wind and energy production exceeds the demands, electrolyzers produce hydrogen for storage and when there is not wind, a hydrogen generator and a fuel cell convert the stored hydrogen back into electricity, supplying a constant power source, autonomous and renewable.

## Small Wind Turbines in the Market

Nowadays, there are tens of wind energy industries involved in the trade of small wind turbines worldwide. Most of these industries offer the consumers more than one model of small wind turbines so that correct selection of a wind turbine, that is appropriate for the consumer's energy needs and application's characteristics, is possible. In the small wind turbine market, one can find both HAWTs and VAWTs (Savonius and Darrieus models) with capacity of a few Watts to some kWatts. Commercial small WTs external parts (mainly the blades) are usually made of plastic material, fiber glass or nylon so that long durability in severe weather is assured. Most of these models are of cut-in wind speed of 2–3 m/sec while the wind speed that they require for nominal power is usually between 10 and 15 m/sec. The models of some companies are designed to stop their function above a certain wind speed range while others have no cut-out wind speed. As far as height, weight and rotor diameter is concerned, they vary from model to model. In bigger WTs, their height can reach values of 30–40 meters, their weight some hundreds of kilos and rotor diameter 12-15 meters. On the other hand, models with smaller height (lower than 15 meters), their weight is usually between 20–30 kilos and their rotor diameter not above 5 meters.

Regardless to these, the vast majority of small wind turbine models are designed to be noiseless and without mechanical vibrations capable of disturbing the habitants of a building where a wind turbine is installed. An automatic start-up of the wind-rotor is usually assured. In addition to this, most of the commercial models are capable of functioning in extreme meteorological conditions like temperatures below −20°C or above +40°C and their lifetime is usually estimated to be 15–20 years. Maintenance is an important issue for the commercial wind turbines but wind energy companies have managed to make their best in this field, many models are maintenance free while others' maintenance is limited to one annual visual inspection. Finally, a lot of interest is given by many companies in the aesthetic part of their models. Some commercial wind turbines, because of their beautiful design and color, can be rightfully considered as works of art, while harmony with the nature and human environment is succeeded.

Small wind turbines that exist out in the market are proposed for various different applications. Most companies point out, as it is natural, a large number of possible applications of their products. Generally, the models are designed to function as an energy source in house electricity demands (like lighting), boats, small telecommunications centers and mobile homes. Environmentally friendly and cost effective, modern small wind turbines can turn out as a very clever solution to the holiday home owners, as they can cover a lot of power needs of their homes like electricity for TV, computers and other piece of equipment. Energy needs of sailboats for battery charging on board keeping navigation lights, inverters and cabin lighting can easily be balanced by the use of a small wind turbine. Certain companies, offer especially designed models for marine applications like sailboats, life rafts, buoys and lighthouses. Caravanners and mobile home owners can enjoy their holidays without irritating and expensive generators, just choosing a model of WT that suits the necessary energy consumption and make them energy independent. There are even portable, lightweight wind turbines that can be carried by one person and produce electricity sufficient to power portable electronic devices such as communication equipment for campers, hunters and border guards. Without the need of grid connection, telecommunications and TV networks can have their remote stations fully powered while water pumping, seems to be also a popular application, advertised by most companies.

Aquaculture and agriculture professionals are another target group for small wind turbine market since with many WT models they can generate power for feeding systems, security systems and small power tools in their farms, no matter how isolated they are. A very interesting subject concerning commercial models is that there are already small wind turbines that can be used in conjunction to solar panels and Diesel generating sets. The importance of the combination of wind power and photovoltaic systems is underlined as hybrid plants are considered to ensure independent energy supply. Hybrid systems provided in the market can be used for weekend houses, mountain huts, pumps, park machines, telecommunication etc. Fuel cell or hydropower systems can be added in some models improving the performance of the whole system and making it more cost-effective. In almost every small wind turbine industry prospectus one can find examples of already installed systems. Applications in cottages, boats, measuring and alarm systems, water pumping, telecommunications centers have shown that power needs in tens of different cases can be covered successfully by these small systems. After all, small WTs' design, efficiency, lifetime, use and maintenance are subjects under continuing amelioration and as years pass, more and more useful applications seem to come out.

## Hybrid Wind/Solar Systems

Photovoltaics and small wind turbines can be effectively combined in many applications. Hybrid (photovoltaic-wind and Diesel) systems can offer great abilities in the production of energy based on wind and solar energy. A Diesel generator is often used so that energy needs are covered in case of insufficient meteorological conditions. A battery can also be used with the hybrid systems for storage of energy when its production overcomes the necessary needs. In regions where sunshine and wind conditions are good, like the Greek islands, the combined use of photovoltaics and wind turbine has great results for most of the day-night period and also for a very large period of a year. A lot of research has been made on the performance of hybrid power systems and experimental results have been published in many articles. The energy output of a hybrid system can be enough for the demands of a house placed in regions where the extension of the already available electricity grid would be financially unadvisable. Such hybrid systems can also be used in various other applications: except from remote and isolated houses, a combined mini-grid of many hybrid systems could be used for energy supply of small communities that wish to follow a more independent and clean way for the coverage of their energy needs.

Hybrid (wind/solar) systems are a very interesting solution for hotels, greenhouses, farms, small telecommunication stations and for any other unit whose location's characteristics give it the ability to exploit, not one, but two alternative energy sources for the generation of a large part of the necessary energy. Of course, before the installation of a hybrid system, a very good study of the climatic characteristics such as the wind potential and the annual solar radiation must certainly be done. The advantages of energy production with a hybrid (wind-solar-Diesel-battery) system are many: energy needs of a unit, for example a house, are covered with the best and most ecological way, Diesel's efficiency reaches maximum while its function damages are very much reduced.

Besides, energy contribution from the Diesel generator and the battery decrease a lot, when the wind turbine and the photovoltaic panels reach the maximum of their performance.

Another type of hybrid system which is proposed as an alternative method of energy production is the photovoltaic/thermal/wind power system. The combination of the PV module with a water or an air heat extraction unit constitutes the hybrid photovoltaic/thermal solar (PVT) solar system, by which electrical and thermal output is simultaneously provided. A PVT/WT hybrid system can cover effectively electrical and thermal needs of a house or a building. Because of the fact that solar radiation and wind vary with time, energy storage is usually necessary to adapt the time of energy conversion with the demand profile. As far as thermal energy is concerned, it is usually stored in water storage tanks for liquid type PVT systems, in rocks (or other material) for air type PVT systems and also in phase change materials (PCM), while electrical energy is mainly stored in batteries. We should not forget to mention that in hybrid systems like PV/WT, the use of hydrogen to store energy is also proposed.

Generally, the performance of these types of hybrid systems, (PV/WT) and (PVT/WT), is continuously under thorough study and research of new, improved models using the combination of photovoltaics and wind turbines is in progress. Results of experiments with hybrid systems have shown that the energy demands may be most cheaply met with a hybrid system than a WT or a PV system alone. This can only lead to better ways of using alternative sources of energy for the electric supply of a load as a domestic use, for the thermal needs of a hotel and for various other applications.

A very important issue that makes hybrid systems a very interesting solution for the production of energy is the complementary function of the photovoltaic panel and the wind turbine. Photovoltaic panels can be useful only in daytime and under a certain solar radiation. On the other hand, WTs can produce energy only when the wind velocity is above a certain rate. So, PV and WT systems can effectively be combined with PVs for sunny days and WTs for windy nights or for the cloudy days. Also, during winter, where the solar radiation is generally at a low rate, PV systems cannot reach a sufficient performance while WTs can offer a lot to energy supply. During summer months, PV systems have fascinating results, something which could compensate the unsteady performance of the WTs.

## Small Wind Turbine Applications in Greece

Wind energy is a scientific domain, which has been left out of the Greek scientific community. Greece is a country with thousands of kilometers of seashores and tens of small and medium islands. With even only these elements, Greece is logically considered as a country where wind energy potential is ideal for many wind energy applications. A lot of research has been done on the development of wind energy in Greece and its future perspectives, including financial issues. Interesting results have been inferred regarding the installation of small wind turbines for energy production in the islands of the Aegean Sea. Among the ideas and the proposals for the use of wind turbines for energy supply are stand alone wind turbine systems, as long as wind-hydro electricity production systems, both of which could cover the energy needs of remote houses, hotels, greenhouses or small communities.

Wind energy potential in Greece, and mostly in the coastal areas and the islands is a very valuable element, which could lead wind energy technology to a road full of perspectives and useful applications. First of all, the use of WTs for the energy needs of remote houses, placed in regions where electricity grid connection would be very difficult, seems to be an ideal and environment friendly solution. According to experiments, if we have a sufficient wind energy potential, a 10 kW wind turbine can usually cover the total or a great part of a household. After all, small WTs, may be combined with Diesel generator so that total energy independence could be assured. Battery could also be used so that when the WT produces energy that overcomes the energy needs, the extra energy could be stored in the battery and be used when the wind conditions would not allow WT generated energy. We must also mention that in most Greek houses, thermal needs as water heating, are covered by electricity. But thermal needs could be instead covered by the use of small wind turbines, something which could be financially competitive.

Hotels in regions where wind velocity is usually of a high level could solve their energy problem by using small wind turbines. A lot of hotels in Greek islands and on the shores of the mainland, have usually a lot of free space in their surroundings which could be used for the installation of one or more WTs of horizontal or vertical axis. Moreover, as it is commonly known, high and well-exposed ground to the wind is the best location for a wind turbine since wind speed increases with height and also over slopes with big inclination. Many hotels in Greece, especially in Greek islands are located in this kind of beneficiary locations: on steep slopes and on well exposed rocky hills. Under these conditions, small electricity needs in big hotels which offer a lot of outdoor activities and facilities, for example the lighting of a tennis court, could easily be covered by small wind turbines that generate the windy days and nights energy, stored in battery in case of excess wind events (high wind or low load intervals). Generally, a detailed research of a hotel's energy consumption in combination with a proper environmental study, so that harmony with the surroundings is maintained, could offer a clever and profitable solution to the hotel's owners. Regardless to this, a hotel which uses alternative energy sources for its energy needs can draw a lot of people's attention and certainly all the ecologists', something which can only turn out as another advantage in the use of WTs for hotels' owners.

Greenhouses and farms in Greece could turn to autonomous units, as far as energy is concerned, with the use of small WTs which could be installed nearby greenhouses and farms. Small WTs could be combined with a Diesel generator for the non windy days and a battery for the storage of extra energy generated. These WTs could be designed especially for each farm or greenhouse so that not only their capacity suits their energy needs, but also, their design and place cannot destroy the overall area's view. Under these circumstances, farmers would find small WTs as a good investment after some years of functioning and with no practical destruction of their land, since a minor percentage of the land area is needed for the wind turbines. But there is another great advantage for the already grid- connected farms and greenhouses. The farmers could sell an amount of energy generated by their WTs and so enhance their income. Not to mention the increase of value of the land owing to the autonomy of the whole property.

Greece is a sea country with thousands of sailing boats. A sailboat's battery charging with electricity by a small wind turbine is a very practical and clever proposal, since wind in the open sea is usually rated over the cut-in wind rate of small wind turbines. In this way, sailboats owners could feel more confident and less dependent to the traditional, expensive ways of electricity supply on their boats.

## Wind-Turbine Aerodynamics

Wind-turbine blades awaiting installation in laydown yard.

The primary application of wind turbines is to generate energy using the wind. Hence, the aerodynamics is a very important aspect of wind turbines. Like most machines, there are many different types of wind turbines, all of them based on different energy extraction concepts.

Though the details of the aerodynamics depend very much on the topology, some fundamental concepts apply to all turbines. Every topology has a maximum power for a given flow, and some topologies are better than others. The method used to extract power has a strong influence on this. In general, all turbines may be grouped as being either lift-based, or drag-based; the former being more efficient. The difference between these groups is the aerodynamic force that is used to extract the energy.

The most common topology is the horizontal-axis wind turbine. It is a lift-based wind turbine with very good performance. Accordingly, it is a popular choice for commercial applications and much research has been applied to this turbine. Despite being a popular lift-based alternative in the latter part of the 20th century, the Darrieus wind turbine is rarely used today. The Savonius wind turbine is the most common drag type turbine. Despite its low efficiency, it remains in use because of its robustness and simplicity to build and maintain.

### General Aerodynamic Considerations

The governing equation for power extraction is stated below:

$$P = \vec{F} \cdot \vec{v}$$

Where: P is the power, F is the force vector, and v is the velocity of the moving wind turbine part.

The force F is generated by the wind's interaction with the blade. The magnitude and distribution of this force is the primary focus of wind-turbine aerodynamics. The most familiar type of aerodynamic force is drag. The direction of the drag force is parallel to the relative wind. Typically, the wind turbine parts are moving, altering the flow around the part. An example of relative wind is the wind one would feel cycling on a calm day.

To extract power, the turbine part must move in the direction of the net force. In the drag force case, the relative wind speed decreases subsequently, and so does the drag force. The relative wind aspect dramatically limits the maximum power that can be extracted by a drag-based wind turbine. Lift-based wind turbines typically have lifting surfaces moving perpendicular to the flow. Here, the relative wind does not decrease; rather, it increases with rotor speed. Thus, the maximum power limits of these machines are much higher than those of drag-based machines.

## Characteristic Parameters

Wind turbines come in a variety of sizes. Once in operation, a wind turbine experiences a wide range of conditions. This variability complicates the comparison of different types of turbines. To deal with this, nondimensionalization is applied to various qualities. Nondimensionalization allows one to make comparisons between different turbines, without having to consider the effect of things like size and wind conditions from the comparison. One of the qualities of nondimensionalization is that though geometrically similar turbines will produce the same non-dimensional results, other factors (difference in scale, wind properties) cause them to produce very different dimensional properties.

The coefficient of power is the most important variable in wind-turbine aerodynamics. The Buckingham $\pi$ theorem can be applied to show that the non-dimensional variable for power is given by the equation below. This equation is similar to efficiency, so values between 0 and less than 1 are typical. However, this is not exactly the same as efficiency and thus in practice, some turbines can exhibit greater than unity power coefficients. In these circumstances, one cannot conclude the first law of thermodynamics is violated because this is not an efficiency term by the strict definition of efficiency.

$$C_P = \frac{P}{\frac{1}{2}\rho A V^3}$$

Where: $C_P$ is the coefficient of power, $\rho$ is the air density, A is the area of the wind turbine, and V is the wind speed.

The thrust coefficient is another important dimensionless number in wind turbine aerodynamics.

$$C_T = \frac{T}{\frac{1}{2}\rho A V^2}$$

Equation $P = \vec{F} \cdot \vec{v}$ shows two important dependents. The first is the speed (U) that the machine is going at. The speed at the tip of the blade is usually used for this purpose, and is written as the product of the blade radius "r" and the rotational speed of the wind ("$U = \Omega r$", where $\Omega$ is the rotational velocity in radians/second).[please clarify] This variable is nondimensionalized by the wind speed, to obtain the speed ratio:

$$\lambda = \frac{U}{V}$$

The force vector is not straightforward, as stated earlier there are two types of aerodynamic forces, lift and drag. Accordingly, there are two non-dimensional parameters. However both variables are non-dimensionalized in a similar way. The formula for lift is given below, the formula for drag is given after:

$$C_L = \frac{L}{\frac{1}{2}\rho A W^2}$$

$$C_D = \frac{D}{\frac{1}{2}\rho A W^2}$$

Where: $C_L$ is the lift coefficient, $C_D$ is the drag coefficient, $W$ is the relative wind as experienced by the wind turbine blade, and A is the area. Note that A may not be the same area used in the power non-dimensionalization of power.

The aerodynamic forces have a dependency on $W$, this speed is the relative speed and it is given by the equation below. Note that this is vector subtraction.

$$\vec{W} = \vec{V} - \vec{U}$$

## Drag- versus Lift-based Machines

All wind turbines extract energy from the wind through aerodynamic forces. There are two important aerodynamic forces: drag and lift. Drag applies a force on the body in the direction of the relative flow, while lift applies a force perpendicular to the relative flow. Many machine topologies could be classified by the primary force used to extract the energy. For example, a Savonious wind turbine is a drag-based machine, while a Darrieus wind turbine and conventional horizontal axis wind turbines are lift-based machines. Drag-based machines are conceptually simple, yet suffer from poor efficiency. Efficiency in this analysis is based on the power extracted vs. the plan-form area. Considering that the wind is free, but the blade materials are not, a plan-form-based definition of efficiency is more appropriate.

The analysis is focused on comparing the maximum power extraction modes and nothing else. Accordingly, several idealizations are made to simplify the analysis, further considerations are required to apply this analysis to real turbines. For example, in this comparison the effects of axial momentum theory are ignored. Axial momentum theory demonstrates how the wind turbine imparts an influence on the wind which in-turn decelerates the flow and limits the maximum power. Since this effect is the same for both lift and drag-based machines it can be ignored for comparison purposes. The topology of the machine can introduce additional losses, for example trailing vorticity in horizontal axis machines degrade the performance at the tip. Typically these losses are minor and can be ignored in this analysis (for example tip loss effects can be reduced with using high aspect-ratio blades).

## Maximum Power of a Drag-based Wind Turbine

Equation ( $P = \vec{F} \cdot \vec{v}$ ) will be the starting point in this derivation. Equation (CD) is used to define the force, and equation (RelativeSpeed) is used for the relative speed. These substitutions give the following formula for power.

$$P = \frac{1}{2}\rho A C_D \left(UV^2 - 2VU^2 + U^3\right)$$

The formulas (CP) and (SpeedRatio) are applied to express (DragPower) in nondimensional form:

$$C_P = C_D \left(\lambda - 2\lambda^2 + \lambda^3\right)$$

It can be shown through calculus that equation (DragCP) achieves a maximum at $\lambda = 1/3$. By inspection one can see that equation (DragPower) will achieve larger values for $\lambda > 1$. In these circumstances, the scalar product in equation ($P = \vec{F} \cdot \vec{v}$) makes the result negative. Thus, one can conclude that the maximum power is given by:

$$C_P = \frac{4}{27} C_D$$

Experimentally it has been determined that a large $C_D$ is 1.2, thus the maximum $C_P$ is approximately 0.1778.

## Maximum Power of a Lift-based Wind Turbine

The derivation for the maximum power of a lift-based machine is similar, with some modifications. First we must recognize that drag is always present, and thus cannot be ignored. It will be shown that neglecting drag leads to a final solution of infinite power. This result is clearly invalid, hence we will proceed with drag. As before, equations ($P = \vec{F} \cdot \vec{v}$), (CD) and (RelativeSpeed) will be used along with (CL) to define the power below expression.

$$P = \frac{1}{2} \rho A \sqrt{U^2 + V^2} \left( C_L UV - C_D U^2 \right)$$

Similarly, this is non-dimensionalized with equations (CP) and (SpeedRatio). However, in this derivation the parameter $\gamma = C_D / C_L$ is also used:

$$C_P = C_L \sqrt{1 + \lambda^2} \left( \lambda - \gamma \lambda^2 \right)$$

Solving the optimal speed ratio is complicated by the dependency on $\gamma$ and the fact that the optimal speed ratio is a solution to a cubic polynomial. Numerical methods can then be applied to determine this solution and the corresponding $C_P$ solution for a range of $\gamma$ results. Some sample solutions are given in the table below.

| $\gamma$ | Optimal $\lambda$ | Optimal $C_P$ |
|---|---|---|
| 0.5 | 1.23 | 0.75 $C_L$ |
| 0.2 | 3.29 | 3.87 $C_L$ |
| 0.1 | 6.64 | 14.98 $C_L$ |
| 0.05 | 13.32 | 59.43 $C_L$ |
| 0.04 | 16.66 | 92.76 $C_L$ |
| 0.03 | 22.2 | 164.78 $C_L$ |
| 0.02 | 33.3 | 370.54 $C_L$ |
| 0.01 | 66.7 | 1481.65 $C_L$ |
| 0.007 | 95.23 | 3023.6 $C_L$ |

Experiments have shown that it is not unreasonable to achieve a drag ratio ($\gamma$) of about 0.01 at a lift coefficient of 0.6. This would give a $C_P$ of about 889. This is substantially better than the best drag-based machine, and explains why lift-based machines are superior.

In the analysis given here, there is an inconsistency compared to typical wind turbine non-dimensionalization. As stated in the preceding section, the A (area) in the $C_P$ non-dimensionalization is not always the same as the A in the force equations (CL) and (CD). Typically for $C_P$ the A is the area swept by the rotor blade in its motion. For $C_L$ and $C_D$ A is the area of the turbine wing section. For drag based machines, these two areas are almost identical so there is little difference. To make the lift based results comparable to the drag results, the area of the wing section was used to non-dimensionalize power. The results here could be interpreted as power per unit of material. Given that the material represents the cost (wind is free), this is a better variable for comparison.

If one were to apply conventional non-dimensionalization, more information on the motion of the blade would be required. However the discussion on Horizontal Axis Wind Turbines will show that the maximum $C_P$ there is 16/27. Thus, even by conventional non-dimensional analysis lift based machines are superior to drag based machines.

There are several idealizations to the analysis. In any lift-based machine (aircraft included) with finite wings, there is a wake that affects the incoming flow and creates induced drag. This phenomenon exists in wind turbines and was neglected in this analysis. Including induced drag requires information specific to the topology, In these cases it is expected that both the optimal speed-ratio and the optimal $C_P$ would be less. The analysis focused on the aerodynamic potential, but neglected structural aspects. In reality most optimal wind-turbine design becomes a compromise between optimal aerodynamic design and optimal structural design.

## Horizontal-axis Wind Turbine

The aerodynamics of a horizontal-axis wind turbine are not straightforward. The air flow at the blades is not the same as the airflow further away from the turbine. The very nature of the way in which energy is extracted from the air also causes air to be deflected by the turbine. In addition, the aerodynamics of a wind turbine at the rotor surface exhibit phenomena rarely seen in other aerodynamic fields.

## Axial Momentum and the Lanchester–Betz–Joukowsky Limit

Energy in fluid is contained in four different forms: gravitational potential energy, thermodynamic pressure, kinetic energy from the velocity and finally thermal energy. Gravitational and thermal energy have a negligible effect on the energy extraction process. From a macroscopic point of view, the air flow about the wind turbine is at atmospheric pressure. If pressure is constant then only kinetic energy is extracted. However up close near the rotor itself the air velocity is constant as it passes through the rotor plane. This is because of conservation of mass. The air that passes through the rotor cannot slow down because it needs to stay out of the way of the air behind it. So at the rotor the energy is extracted by a pressure drop. The air directly behind the wind turbine is at sub-atmospheric pressure; the air in front is under greater than atmospheric pressure. It is this high pressure in front of the wind turbine that deflects some of the upstream air around the turbine.

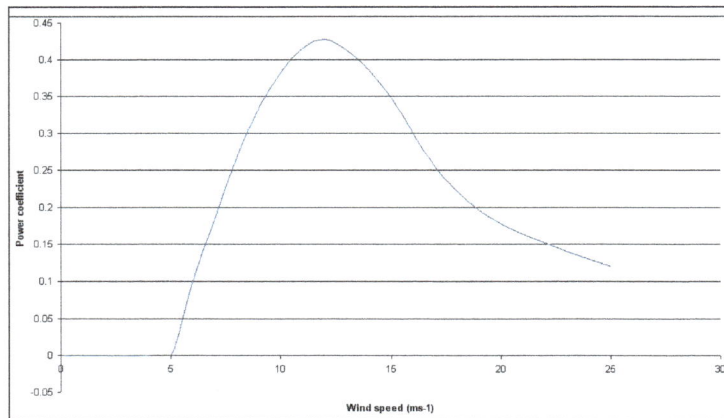

Wind turbine power coefficient.

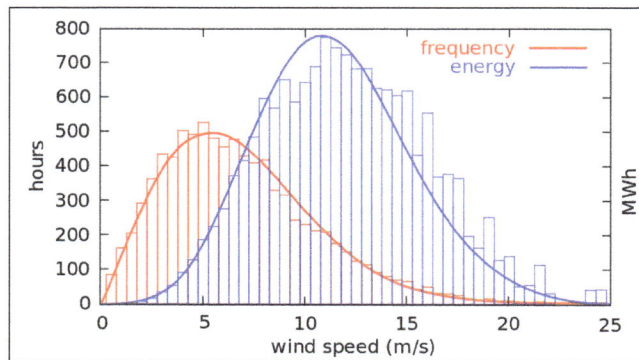

Distribution of wind speed (red) and energy generated (blue). The histogram
shows measured data, while the curve is the Rayleigh model distribution for the same average wind speed.

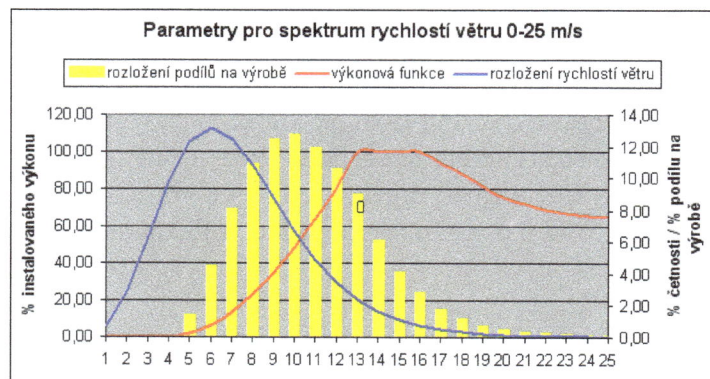

Distribution of wind speed (blue) and energy generated (yellow).

Frederick W. Lanchester was the first to study this phenomenon in application to ship propellers, five years later Nikolai Yegorovich Zhukovsky and Albert Betz independently arrived at the same results. It is believed that each researcher was not aware of the others' work because of World War I and the Bolshevik Revolution. Thus formally, the proceeding limit should be referred to as the Lanchester–Betz–Joukowsky limit.

This is derived by looking at the axial momentum of the air passing through the wind turbine. As stated above some of the air is deflected away from the turbine. This causes the air passing through the rotor plane to have a smaller velocity than the free stream velocity. The ratio of this reduction to that of the air velocity far away from the wind turbine is called the axial induction factor. It is defined as below:

$$a \equiv \frac{U_1 - U_2}{U_1}$$

where a is the axial induction factor, $U_1$ is the wind speed far away upstream from the rotor, and $U_2$ is the wind speed at the rotor.

The first step to deriving the Betz limit is applying conservation of angular momentum. As stated above the wind loses speed after the wind turbine compared to the speed far away from the turbine. This would violate the conservation of momentum if the wind turbine was not applying a thrust force on the flow. This thrust force manifests itself through the pressure drop across the rotor. The front operates at high pressure while the back operates at low pressure. The pressure difference from the front to back causes the thrust force. The momentum lost in the turbine is balanced by the thrust force.

Another equation is needed to relate the pressure difference to the velocity of the flow near the turbine. Here the Bernoulli equation is used between the field flow and the flow near the wind turbine. There is one limitation to the Bernoulli equation: the equation cannot be applied to fluid passing through the wind turbine. Instead conservation of mass is used to relate the incoming air to the outlet air. Betz used these equations and managed to solve the velocities of the flow in the far wake and near the wind turbine in terms of the far field flow and the axial induction factor. The velocities are given below as:

$$U_2 = U_1(1-a)$$
$$U_4 = U_1(1-2a)$$

$U_4$ is introduced here as the wind velocity in the far wake. This is important because the power extracted from the turbine is defined by the following equation. However the Betz limit is given in terms of the coefficient of power $C_p$. The coefficient of power is similar to efficiency but not the same. The formula for the coefficient of power is given beneath the formula for power:

$$P = 0.5\rho A U_2(U_1^2 - U_4^2)$$
$$C_p \equiv \frac{P}{0.5\rho A U_1^3}$$

Betz was able to develop an expression for $C_p$ in terms of the induction factors. This is done by the velocity relations being substituted into power and power is substituted into the coefficient of power definition. The relationship Betz developed is given below:

$$C_p = 4a(1-a)^2$$

The Betz limit is defined by the maximum value that can be given by the above formula. This is found by taking the derivative with respect to the axial induction factor, setting it to zero and solving for the axial induction factor. Betz was able to show that the optimum axial induction factor is one third. The optimum axial induction factor was then used to find the maximum coefficient of power. This maximum coefficient is the Betz limit. Betz was able to show that the maximum coefficient of power of a wind turbine is 16/27. Airflow operating at higher thrust will cause the axial induction factor to rise above the optimum value. Higher thrust cause more air to be deflected away from the turbine. When the axial induction factor falls below the optimum value the wind turbine is not extracting all the energy it can. This reduces pressure around the turbine and allows more air to pass through the turbine, but not enough to account for the lack of energy being extracted.

The derivation of the Betz limit shows a simple analysis of wind turbine aerodynamics. In reality there is a lot more. A more rigorous analysis would include wake rotation, the effect of variable geometry. The effect of airfoils on the flow is a major component of wind turbine aerodynamics. Within airfoils alone, the wind turbine aerodynamicist has to consider the effect of surface roughness, dynamic stall tip losses, solidity, among other problems.

## Angular Momentum and Wake Rotation

The wind turbine described by Betz does not actually exist. It is merely an idealized wind turbine described as an actuator disk. It's a disk in space where fluid energy is simply extracted from the air. In the Betz turbine the energy extraction manifests itself through thrust. The equivalent turbine described by Betz would be a horizontal propeller type operating at an infinite tip speed ratios and no losses. The tip speed ratio is ratio of the speed of the tip relative to the free stream flow. Actual turbines try to run very high L/D airfoils at high tip speed ratios to attempt to approximate this, but there are still additional losses in the wake because of these limitations.

One key difference between actual turbines and the actuator disk, is that the energy is extracted through torque. The wind imparts a torque on the wind turbine, thrust is a necessary by-product of torque. Newtonian physics dictates that for every action there is an equal and opposite reaction. If the wind imparts a torque on the blades then the blades must be imparting a torque on the wind. This torque would then cause the flow to rotate. Thus the flow in the wake has two components, axial and tangential. This tangential flow is referred to as wake rotation.

Torque is necessary for energy extraction. However wake rotation is considered a loss. Accelerating the flow in the tangential direction increases the absolute velocity. This in turn increases the amount of kinetic energy in the near wake. This rotational energy is not dissipated in any form that would allow for a greater pressure drop (Energy extraction). Thus any rotational energy in the wake is energy that is lost and unavailable.

This loss is minimized by allowing the rotor to rotate very quickly. To the observer it may seem like the rotor is not moving fast; however, it is common for the tips to be moving through the air at 8-10 times the speed of the free stream. Newtonian mechanics defines power as torque multiplied by the rotational speed. The same amount of power can be extracted by allowing the rotor to rotate faster and produce less torque. Less torque means that there is less wake rotation. Less wake rotation means there is more energy available to extract. However, very high tip speeds also increase the drag on the blades, decreasing power production. Balancing these factors is what leads to most modern horizontal axis wind turbines running at a tip speed ratio around 9. In addition, wind turbines usually limit the tip speed to around 80-90m/s due to leading edge erosion and high noise levels. At wind speeds above about 10m/s (where a turbine running a tip speed ratio of 9 would reach 90m/s tip speed), turbines usually do not continue to increase rotational speed for this reason, which slightly reduces efficiency.

## Blade Element and Momentum Theory

The simplest model for horizontal axis wind turbine aerodynamics is blade element momentum theory. The theory is based on the assumption that the flow at a given annulus does not affect the flow at adjacent annuli. This allows the rotor blade to be analyzed in sections, where the resulting forces are summed over all sections to get the overall forces of the rotor. The theory uses both axial and angular momentum balances to determine the flow and the resulting forces at the blade.

The momentum equations for the far field flow dictate that the thrust and torque will induce a secondary flow in the approaching wind. This in turn affects the flow geometry at the blade. The blade itself is the source of these thrust and torque forces. The force response of the blades is governed by the geometry of the flow, or better known as the angle of attack. This interplay between the far field momentum balances and the local blade forces requires one to solve the momentum equations and the airfoil equations simultaneously. Typically computers and numerical methods are employed to solve these models.

There is a lot of variation between different versions of blade element momentum theory. First, one can consider the effect of wake rotation or not. Second, one can go further and consider the pressure drop induced in wake rotation. Third, the tangential induction factors can be solved with a momentum equation, an energy balance or orthogonal geometric constraint; the latter a result of Biot–Savart law in vortex methods. These all lead to different set of equations that need to be solved. The simplest and most widely used equations are those that consider wake rotation with the momentum equation but ignore the pressure drop from wake rotation. Those equations are given below. $a$ is the axial component of the induced flow, $a'$ is the tangential component of the induced flow. $\sigma$ is the solidity of the rotor, $\phi$ is the local inflow angle. $C_n$ and $C_t$ are the coefficient of normal force and the coefficient of tangential force respectively. Both these coefficients are defined with the resulting lift and drag coefficients of the airfoil:

$$a = \frac{1}{\dfrac{4}{C_n \sigma} \sin^2 \phi + 1}$$

$$a' = \frac{1}{\dfrac{4}{C_t \sigma} \sin \phi \cos \phi - 1}$$

## Corrections to Blade Element Momentum Theory

Blade element momentum theory alone fails to represent accurately the true physics of real wind turbines. Two major shortcomings are the effects of a discrete number of blades and far field effects when the turbine is heavily loaded. Secondary shortcomings originate from having to deal with transient effects like dynamic stall, rotational effects like the Coriolis force and centrifugal pumping, and geometric effects that arise from coned and yawed rotors. The current state of the art in blade element momentum theory uses corrections to deal with these major shortcomings. These corrections are discussed below. There is as yet no accepted treatment for the secondary shortcomings. These areas remain a highly active area of research in wind turbine aerodynamics.

The effect of the discrete number of blades is dealt with by applying the Prandtl tip loss factor. The most common form of this factor is given below where B is the number of blades, R is the outer radius and r is the local radius. The definition of F is based on actuator disk models and not directly applicable to blade element momentum theory. However the most common application multiplies induced velocity term by F in the momentum equations. As in the momentum equation there are many variations for applying F, some argue that the mass flow should be corrected in either the axial equation, or both axial and tangential equations. Others have suggested a second tip loss term to account for the reduced blade forces at the tip. Shown below are the above momentum equations with the most common application of F:

$$F = \frac{2}{\pi} \arccos \left[ e^{-\frac{B(R-r)}{2r\sin\phi}} \right]$$

$$a = \frac{1}{\frac{4}{C_n \sigma} F \sin^2 \phi + 1}$$

$$a' = \frac{1}{\frac{4}{C_t \sigma} F \sin \phi \cos \phi - 1}$$

The typical momentum theory is effective only for axial induction factors up to 0.4 (thrust coefficient of 0.96). Beyond this point the wake collapses and turbulent mixing occurs. This state is highly transient and largely unpredictable by theoretical means. Accordingly, several empirical relations have been developed. As the usual case there are several version, however a simple one that is commonly used is a linear curve fit given below, with $a_c = 0.2$. The turbulent wake function given excludes the tip loss function, however the tip loss is applied simply by multiplying the resulting axial induction by the tip loss function.

$$C_T = 4\left[ a_c^2 + (1-2a_c)a \right] \text{when } a > a_c$$

The terms $C_T$ and $C_t$ represent different quantities. The first one is the thrust coefficient of the rotor, which is the one which should be corrected for high rotor loading (i.e., for high values of $a$), while the second one ($c_t$) is the tangential aerodynamic coefficient of an individual blade element, which is given by the aerodynamic lift and drag coefficients.

## Aerodynamic Modeling

Blade element momentum theory is widely used due to its simplicity and overall accuracy, but its originating assumptions limit its use when the rotor disk is yawed, or when other non-axisymmetric effects (like the rotor wake) influence the flow. Limited success at improving predictive accuracy has been made using computational fluid dynamics (CFD) solvers based on Reynolds-averaged Navier–Stokes equations and other similar three-dimensional models such as free vortex methods. These are very computationally intensive simulations to perform for several reasons. First, the solver must accurately model the far-field flow conditions, which can extend several rotor diameters up- and down-stream and include atmospheric boundary layer turbulence, while at the same time resolving the small-scale boundary-layer flow conditions at the blades' surface (necessary to capture blade stall). In addition, many CFD solvers have difficulty meshing parts that move and deform, such as the rotor blades. Finally, there are many dynamic flow phenomena that are not easily modelled by Reynolds-averaged Navier–Stokes equations, such as dynamic stall and tower shadow. Due to the computational complexity, it is not currently practical to use these advanced methods for wind turbine design, though research continues in these and other areas related to helicopter and wind turbine aerodynamics.

Free vortex models and Lagrangian particle vortex methods are both active areas of research that seek to increase modelling accuracy by accounting for more of the three-dimensional and unsteady flow effects than either blade element momentum theory or Reynolds-averaged Navier–Stokes equations. Free vortex models are similar to lifting line theory in that it assumes that the wind turbine rotor is shedding either a continuous vortex filament from the blade tips (and often the root), or a continuous vortex sheet from the blades' trailing edges. Lagrangian particle vortex methods can use a variety of methods to introduce vorticity into the wake. Biot–Savart summation is used to determine the induced flow field of these wake vortices' circulations, allowing for better approximations of the local flow over the rotor blades. These methods have largely confirmed much of the applicability of blade element momentum theory and shed insight into the structure of wind turbine wakes. Free vortex models have limitations due to its origin in potential flow theory, such as not explicitly modelling model viscous behavior (without semi-empirical core models), though the Lagrangian particle vortex methods is a fully viscous method. Lagrangian particle vortex methods are more computationally intensive than either free vortex models or Reynolds-averaged Navier–Stokes equations, and free vortex models still rely on blade element theory for the blade forces.

## Variable Speed Wind Turbine

A variable speed wind turbine is one which is specifically designed to operate over a wide range of rotor speeds. It is in direct contrast to fixed speed wind turbine where the rotor speed is approximately constant. The reason to vary the rotor speed is to capture the maximum aerodynamic power in the wind, as the wind speed varies. The aerodynamic efficiency, or coefficient of power, $C_p$ for a fixed blade pitch angle is obtained by operating the wind turbine at the optimal tip-speed ratio as shown in the following graph.

Tip-speed ratio is given by the following expression,

$$\lambda = \frac{\omega R}{v}$$

where $\omega$ is the rotor speed (in radians per second), $R$ is the radius of the rotor, and $v$ is the wind speed. As the wind speed varies, the rotor speed must be varied to maintain peak efficiency.

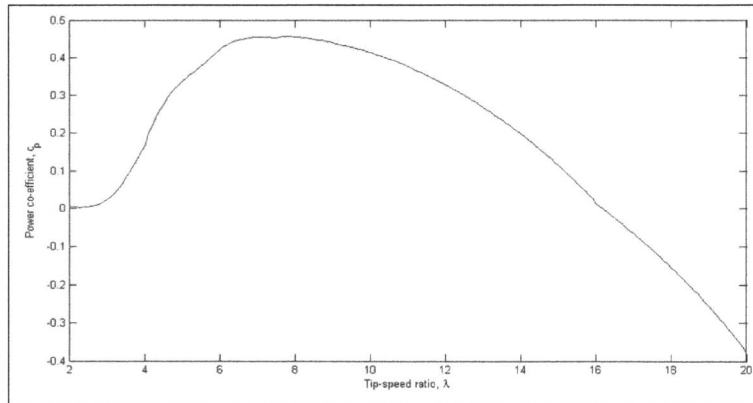

Before the need to connect wind turbines to the grid, turbines were fixed-speed. This was not a problem because turbines did not have to be synchronized with the frequency of the grid.

All grid-connected wind turbines, from the first one in 1939 until the development of variable-speed grid-connected wind turbines in the 1970s, were fixed-speed wind turbines. As of 2003, nearly all grid-connected wind turbines operate at an exactly constant speed (synchronous generators) or within a few percents of constant speed (induction generators).

## Torque Rotor-speed Diagrams

For a wind turbine, the power harvested is given by the following formula:

$$P = \frac{1}{2} \rho \pi R^2 v^3 C_p(\lambda)$$

where $P$ is the aerodynamic power and $\rho$ is the density of the air. The power coefficient is a representation of how much of the available power in the wind is captured by the wind turbine and can be looked up in the graph above.

The torque, $Q$, on the rotor shaft is given by the ratio of the power extracted to the rotor speed:

$$Q = \frac{P}{\omega}$$

Thus we can get the following expressions for torque and power:

$$P = \frac{1}{2\lambda^3} \rho \pi R^5 \omega^3 C_p(\lambda)$$

and, $\quad Q = \frac{1}{2\lambda^3} \rho \pi R^5 \omega^2 C_p(\lambda) = \frac{1}{2\lambda} \rho \pi R^3 v^2 C_p(\lambda)$

From the above equation, we can construct a torque-speed diagram for a wind turbine. This consists of multiple curves: a constant power curve which plots the relationship between torque and rotor speed for constant power (green curve); constant wind speed curves, which plot the relationship between torque and rotor speed for constant wind speeds (dashed grey curves); and constant efficiency curves, which plot the relationship between torque and rotor speed for constant efficiencies, $C_p$. This diagram is presented below:

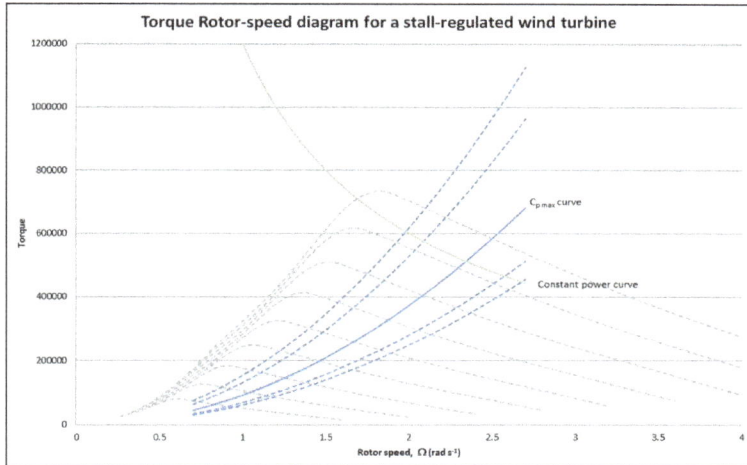

Green curve: Plot of power = rated power so that $P = Q\omega$.

Grey curve: Wind speed is assumed constant so that $Q \propto \omega^2 C_p(\lambda)$.

Blue curve: Constant $C_p(\lambda)$ so that $Q \propto \omega^2$.

## Blade Forces

Consider the following figure:

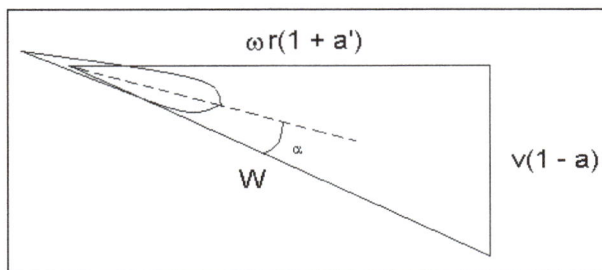

This is the depiction of the apparent wind speed, as seen by a blade. The apparent wind speed is influenced by both the free-stream velocity of the air, and the rotor speed. From this figure, we can see that both the angle $\theta$ and the apparent wind speed $W$ are functions of the rotor speed, $\omega$. By extension, the lift and drag forces will also be functions of $\omega$. This means that the axial and tangential forces that act on the blade vary with rotor speed.

## Operating Strategies for Variable Speed Wind Turbines

### Stall Regulated

A wind turbine would ideally operate at its maximum efficiency for below rated power. Once rated power has been hit, the power is limited. This is for two reasons: ratings on the drivetrain equipment, such as the generator; and second to reduce the loads on the blades. An operating strategy for a wind turbine can thus be divided into a sub-rated-power component, and a rated-power component.

## Below Rated Power

Below rated power, the wind turbine will ideally operate in such a way that $C_p = C_{p\,max}$. On a Torque-rotor speed diagram, this looks as follows:

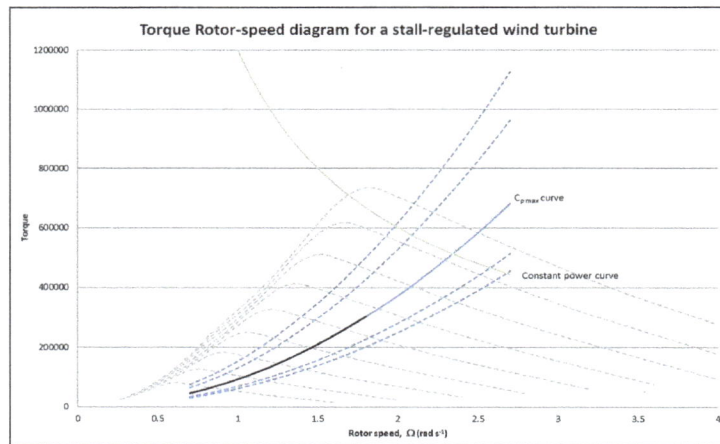

where the black line represents the initial section of the operating strategy for a variable speed stall-regulated wind turbine. Ideally, we would want to stay on the maximum efficiency curve until rated power is hit. However, as the rotor speed increases, the noise levels increase. To counter this, the rotor speed is not allowed to increase above a certain value. This is illustrated in the figure below:

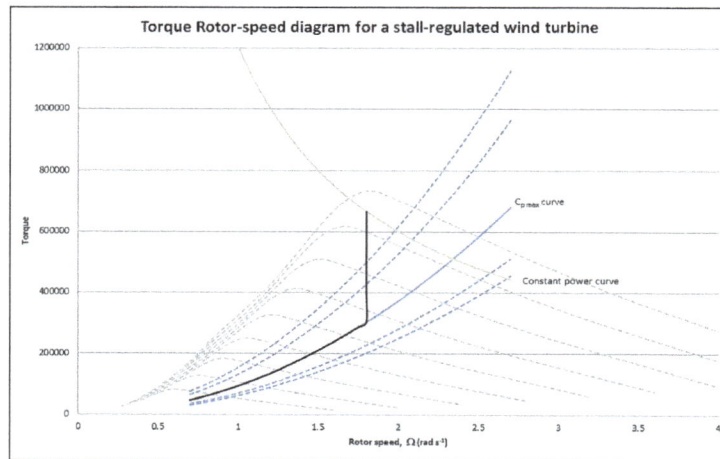

## Rated Power and Above

Once the wind speed has reached a certain level, called rated wind speed, the turbine should not be able to produce any greater levels of power for higher wind speeds. A stall-regulated variable speed wind turbine has no pitching mechanism. However, the rotor speed is variable. The rotor speed can either be increased or decreased by an appropriately designed controller. In reference to the figure illustrated in the blade forces section, it is evident that the angle between the apparent wind speed and the plane of rotation is dependent upon the rotor speed. This angle is termed the angle of attack.

The lift and drag co-efficients for an airfoil are related to the angle of attack. Specifically, for high angles of attack, an airfoil stalls. That is, the drag substantially increases. The lift and drag forces influence the power production of a wind turbine. This can be seen from an analysis of the forces acting on a blade as air interacts with the blade. Thus, forcing the airfoil to stall can result in power limiting.

So it can be established that if the angle of attack needs to be increased to limit the power production of the wind turbine, the rotor speed must be reduced. Again, this can be seen from the figure in the blade forces section. It can also be seen from considering the torque-rotor speed diagram. In reference to the above torque-rotor speed diagram, by reducing the rotor speed at high wind speeds, the turbine enters the stall region, thus bringing some limiting to the power output.

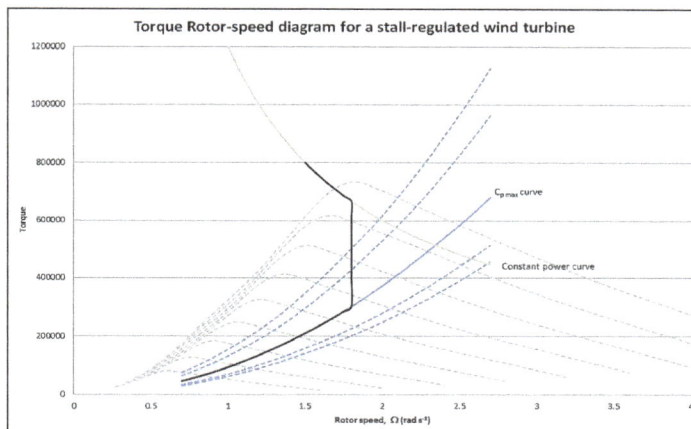

Torque Rotor-speed diagram for a stall-regulated wind turbine

## Pitch Regulated

Pitch regulation thus allows the wind turbine to actively change the angle of attack of the air on the blades. This is preferred over a stall-regulated wind turbine as it enables far greater control of the power output.

## Below Rated Power

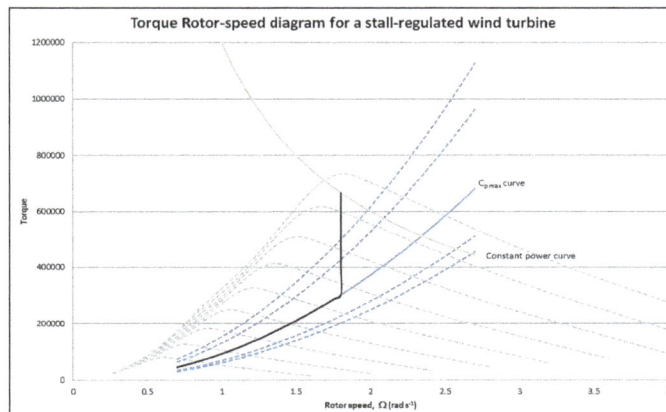

Torque Rotor-speed diagram for a stall-regulated wind turbine

Identical to the stall-regulated variable-speed wind turbine, the initial operating strategy is to operate on the $C_{p\,max}$ curve. However, due to constraints such as noise levels, this is not possible for the full range of sub-rated wind speeds. Below the rated wind speed, the following operating strategy is employed:

## Above Rated Power

Above the rated wind speed, the pitching mechanism is employed. This allows a good level of control over the angle of attack, thus control over the torque. The previous torque rotor-speed diagrams are all plots when the pitch angle, $\beta$, is zero. A three dimensional plot can be produced which includes variations in pitch angle.

Ultimately, in the 2D plot, above rated wind speed, the turbine will operate at the point marked 'x' on the diagram below.

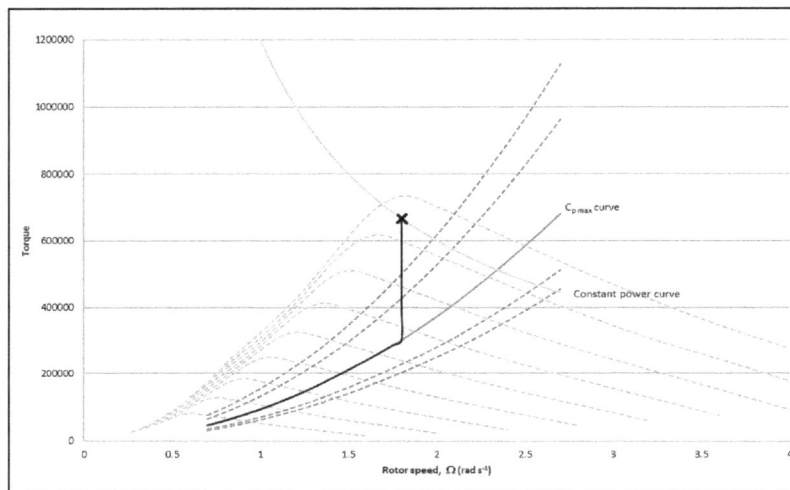

## Gearboxes

A variable speed may or may not have a gearbox, depending on the manufacturer's desires. Wind turbines without gearboxes are called direct-drive wind turbines. An advantage of a gearbox is that generators are typically designed to have the rotor rotating at a high speed within the stator. Direct drive wind turbines do not exhibit this feature. A disadvantage of a gearbox is reliability and failure rates.

An example of a wind turbine without a gearbox is the Enercon E82.

## Generators

For variable speed wind turbines, one of two types of generators can be used: a DFIG (doubly fed induction generator) or an FRC (fully rated converter).

A DFIG generator draws reactive power from the transmission system; this can increase the vulnerability of a transmission system in the event of a failure. A DFIG configuration will require the generator to be a wound rotor; squirrel cage rotors cannot be used for such a configuration.

A fully rated converter can either be an induction generator or a permanent magnet generator. Unlike the DFIG, the FRC can employ a squirrel cage rotor in the generator; an example of this is the Siemens SWT 3.6-107, which is termed the industry workhorse. An example of a permanent magnet generator is the Siemens SWT-2.3-113. A disadvantage of a permanent magnet generator is the cost of materials that need to be included.

## Grid Connections

Consider a variable speed wind turbine with a permanent magnet synchronous generator. The generator produces AC electricity. The frequency of the AC voltage generated by the wind turbine is a function of the speed of the rotor within the generator:

$$N = \frac{120 f}{P}$$

where $N$ is the rotor speed, $P$ is the number of poles in the generator, and $f$ is the frequency of the output Voltage. That is, as the wind speed varies, the rotor speed varies, and so the frequency of the Voltage varies. This form of electricity cannot be directly connected to a transmission system. Instead, it must be corrected such that its frequency is constant. For this, power converters are employed, which results in the de-coupling of the wind turbine from the transmission system. As more wind turbines are included in a national power system, the inertia is decreased. This means that the frequency of the transmission system is more strongly affected by the loss of a single generating unit.

## Power Converters

As already mentioned, the voltage generated by a variable speed wind turbine is non-grid compliant. In order to supply the transmission network with power from these turbines, the signal must be passed through a power converter, which ensures that the frequency of the voltage of the electricity being generated by the wind turbine is the frequency of the transmission system when it is transferred onto the transmission system. Power converters first convert the signal to DC, and then convert the DC signal to an AC signal. Techniques used include pulse width modulation.

## Types of Wind Turbine

Basically the turbines are used to convert the wind energy into electrical energy with the help of a generator. It extracts the energy from the wind and converts it into mechanical energy and then this mechanical energy is used to derive a generator and we get electricity. So let's begin our journey to learn about types of wind turbines.

On the basis of axis of rotation of the blades, it is divided into two parts:

- Horizontal axis wind turbine (HAWT).

- Vertical axis wind turbine (VAWT).

## Horizontal Axis Wind Turbine (HAWT)

It is a turbine in which the axis of rotation of rotor is parallel to the ground and also parallel to wind direction.

They are further divided into two types:

- Upwind turbine.

- Downwind turbine.

- Upwind Turbine.

The turbine in which the rotor faces the wind first are called upwind turbine:

- Today most of the HAWT is manufactured with this design.

- This turbine must be inflexible and placed at some distance from the tower.

- The basic advantage of this turbine is that, it is capable of avoiding wind shade behind the tower.

- It requires yaw mechanism, so that its rotor always faces the wind.

- Downwind Turbine:

  ◦ The turbine in which the rotor is present at the downside of the tower is called downwind turbine. In these types of wind turbines, the wind first faces the tower and after that it faces the rotor blades.

  ◦ Yaw mechanism is absent in this turbine. The rotors and nacelles are designed in such a way that the nacelle allows the wind to flow in a controlled manner.

  ◦ It receives some fluctuation in wind power because here the rotor passes through the wind shade of the tower. In other words the rotor is present after nacelle of the tower and this create fluctuation in the wind power.

## Advantages and Disadvantages of HAWTs

The various advantages and disadvantages of the horizontal axis types of wind turbines are:

## Advantages

- It has self-starting ability. It does not require any external power source to start.

- It has high efficiency as compared with the HAWT.

- Capable of working in high wind speed condition.

- In the case of slow wind condition, its angle of attack can be varied to get maximum possible efficiency.

- Since all blades of this turbine work simultaneously, so it is capable of extracting maximum energy form the wind.

## Disadvantages

- Its initial installation cost is high.

- It requires large ground area for its installation.

- Because of its giant size of blades and towers, it becomes difficult to transport it to the sites.

- High maintenance cost.

- Creates noise problem.

- It cannot be installed near human population.

- It is not good for the bird's population. They are killed by its blades rotation.

## Vertical Axis Wind Turbine (VAWT)

It is a turbine in which the axis of rotation of the rotor is perpendicular to the ground and also perpendicular to the wind direction.

- It can operates in low wind situation.

- It is easier to build and transport.

- These types of Wind turbines are mounted close to the ground and are capable of handling turbulence in far better way as compared with the HAWT.

- Because of its less efficiency, it is used only for the private purpose.

VAWTs are further classified as:

- Darrieus turbine.

- Giromill turbine.

- Savonius turbine.

- Darrieus turbine.

Darrieus turbine is type of HAWT. It was first discovered and patented in 1931 by French aeronautical engineer, Georges Jean Marie Darrieus. It is also known as egg beater turbine because of its egg beater shaped rotor blades.

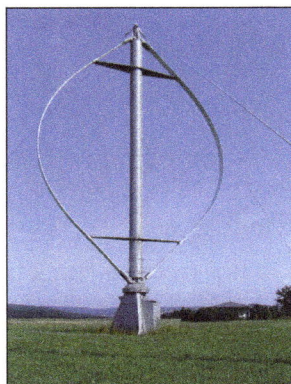

- It consists of vertically oriented blades which are mounted on a vertical rotor. It is not a self-starting turbine and hence a small powered motor is required to start its rotation.

- First the Darrieus turbine is rotated by using a small powered motor. Once it attains sufficient speed, the wind flowing across its blades generates lift forces and this lift forces provides the necessary torque for the rotation. As the rotor rotates, it also rotates the generator and electricity is produced.

- Giromill Turbine.

- It is similar to the Darrieus turbine but the difference is that, it has H-shaped rotor. It works on the same principle of Darrieus turbine:

  ◦ This turbine has H- shaped rotor. Here Darrieus design which has egg beater shaped rotor blades are replaced by straight vertical blades attached with central tower with horizontal supports. It may consists of 2-3 rotor blades.

  ◦ Giromill turbine is cheap and easy to build as compared with Darrieus turbine. It is less efficient turbine and requires strong wind to start. Same as darrieus types of wind turbines, it is also not self- starting and requires small powered motor to start. It is capable of working in turbulent wind conditions.

- Savonius Turbine.

- Savonius turbine is HAWT. It was first discovered in 1922 by a Finnish Engineer Sigurd Johannes Savonius. It is one of the simplest turbine among all known turbines:

  ◦ It is a drag-type device and consists of two or three scoops. If we look it from above than it looks 'S' shape in cross section. The scoops of these turbines have curvature shape and because of that, it experiences less drag when it moves against the wind instead of moving with the wind.

  ◦ Since it is a drag-type machine, it is capable of extracting very less amount of wind power as compared with other similar sized lift-type turbines.

## Advantages and Disadvantages of Vertical Axis Wind Turbine

### Advantages

- It is simple in design and easy to construct and transport.

- It can be easily installed to desired location.

- It requires less ground area for its installation.

- Initial installation cost is very less as compared with the HAWT.

- It can work in turbulent wind condition.

- It is omni-directional and hence do not need to track winds.

- They are smaller in size and hence can be used for domestic or private purpose easily.

- They have low maintenance cost as compared with the HAWT.

### Disadvantages

- It is less efficient. The efficiency of this turbine is about 30-35%.

- They are not self-starting. A small powered motor is needed to start it.

- Guy wires may required to support this turbine.

### Difference between Horizontal Axis Wind Turbine and Vertical Axis Wind Turbine

The various difference between horizontal axis and vertical axis types of wind turbines in tabular form are given below:

| Horizontal Axis Wind Turbine | Vertical Axis Wind Turbine |
|---|---|
| In HAWTs, the axis of rotation of the rotor is Horizontal to the ground. | In VAWTs the axis of rotation of the rotor is perpendicular to the ground. |
| Yaw mechanism is present. | Absence of Yaw mechanism. |
| It has high initial installation cost. | It has low initial installation cost. |
| They are big in size. | They are small in size. |
| Its efficiency is high. | It has low efficiency. |
| It requires large ground area for installation. | It requires less ground area for installation. |
| High maintenance cost. | Low maintenance cost as compared with HAWT. |
| They are self-starting. | They are not self-starting. |
| They are unable to work in low wind speed condition. | They are capable of working in low wind speed condition. |
| Difficult in transportation. | Easy in transportation. |
| They are mostly used commercially. | They are mostly used for private purpose only. |
| It cannot be installed near human population. | It can be installed near human population. |
| It is not good for the bird's population. | It is good for the bird's population. |

## Darrieus Wind Turbine

The Darrieus wind turbine is a type of vertical axis wind turbine (VAWT) used to generate electricity from wind energy. The turbine consists of a number of curved aerofoil blades mounted on a rotating shaft or framework. The curvature of the blades allows the blade to be stressed only in tension at high rotating speeds. There are several closely related wind turbines that use straight blades. This design of the turbine was patented by Georges Jean Marie Darrieus, a French aeronautical engineer; filing for the patent was October 1, 1926. There are major difficulties in protecting the Darrieus turbine from extreme wind conditions and in making it self-starting.

A Darrieus wind turbine once used to generate electricity on the Magdalen Islands.

## Method of Operation

In the original versions of the Darrieus design, the aerofoils are arranged so that they are symmetrical and have zero rigging angle, that is, the angle that the aerofoils are set relative to the structure on which they are mounted. This arrangement is equally effective no matter which direction the wind is blowing—in contrast to the conventional type, which must be rotated to face into the wind.

Combined Darrieus–Savonius generator used in Taiwan.

When the Darrieus rotor is spinning, the aerofoils are moving forward through the air in a circular path. Relative to the blade, this oncoming airflow is added vectorially to the wind, so that the resultant airflow creates a varying small positive angle of attack to the blade. This generates a net force pointing obliquely forwards along a certain 'line-of-action'. This force can be projected inwards past the turbine axis at a certain distance, giving a positive torque to the shaft, thus helping it to rotate in the direction it is already travelling in. The aerodynamic principles which rotate the rotor are equivalent to that in autogiros, and normal helicopters in autorotation.

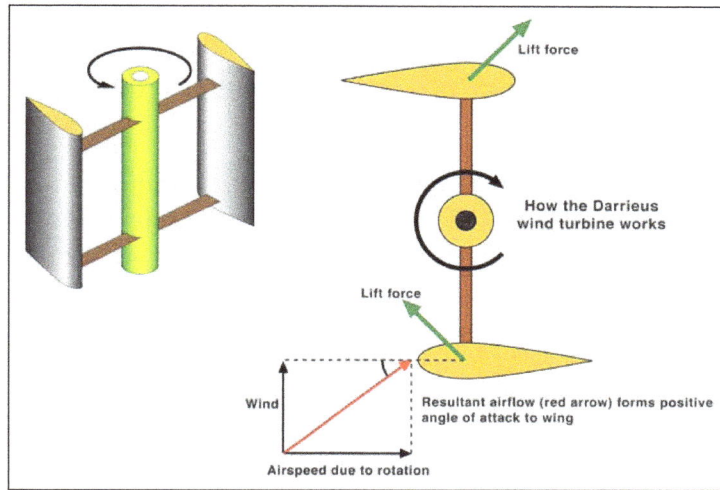

How the Darrieus wind turbine works.

As the aerofoil moves around the back of the apparatus, the angle of attack changes to the opposite sign, but the generated force is still obliquely in the direction of rotation, because the wings are symmetrical and the rigging angle is zero. The rotor spins at a rate unrelated to the windspeed, and usually many times faster. The energy arising from the torque and speed may be extracted and converted into useful power by using an electrical generator.

The aeronautical terms lift and drag are, strictly speaking, forces across and along the approaching net relative airflow respectively, so they are not useful here. We really want to know the tangential force pulling the blade around, and the radial force acting against the bearings.

When the rotor is stationary, no net rotational force arises, even if the wind speed rises quite high—the rotor must already be spinning to generate torque. Thus the design is not normally self-starting. Under rare conditions, Darrieus rotors can self-start, so some form of brake is required to hold it when stopped.

One problem with the design is that the angle of attack changes as the turbine spins, so each blade generates its maximum torque at two points on its cycle (front and back of the turbine). This leads to a sinusoidal (pulsing) power cycle that complicates design. In particular, almost all Darrieus turbines have resonant modes where, at a particular rotational speed, the pulsing is at a natural frequency of the blades that can cause them to (eventually) break. For this reason, most Darrieus turbines have mechanical brakes or other speed control devices to keep the turbine from spinning at these speeds for any lengthy period of time.

Another problem arises because the majority of the mass of the rotating mechanism is at the periphery rather than at the hub, as it is with a propeller. This leads to very high centrifugal stresses on the mechanism, which must be stronger and heavier than otherwise to withstand them. One common approach to minimise this is to curve the wings into an "egg-beater" shape (this is called a "troposkein" shape, derived from the Greek for "the shape of a spun rope") such that they are self-supporting and do not require such heavy supports and mountings.

In this configuration, the Darrieus design is theoretically less expensive than a conventional type, as most of the stress is in the blades which torque against the generator located at the bottom of the turbine. The only forces that need to be balanced out vertically are the compression load due to the blades flexing outward (thus attempting to "squeeze" the tower), and the wind force trying to blow the whole turbine over, half of which is transmitted to the bottom and the other half of which can easily be offset with guy wires.

By contrast, a conventional design has all of the force of the wind attempting to push the tower over at the top, where the main bearing is located. Additionally, one cannot easily use guy wires to offset this load, because the propeller spins both above and below the top of the tower. Thus the conventional design requires a strong tower that grows dramatically with the size of the propeller. Modern designs can compensate most tower loads of that variable speed and variable pitch.

In overall comparison, while there are some advantages in Darrieus design there are many more disadvantages, especially with bigger machines in the MW class. The Darrieus design uses much more expensive material in blades while most of the blade is too close to the ground to give any real power. Traditional designs assume that the wing tip is at least 40 m from ground at lowest point to maximize energy production and lifetime. So far there is no known material (not even carbon fiber) which can meet cyclic load requirements.

## Giromills

A Giromill-type wind turbine.

MUCE turbines installed atop the Marine Board Building.

Darrieus's 1927 patent also covered practically any possible arrangement using vertical airfoils. One of the more common types is the H-rotor, also called the Giromill or H-bar design, in which the long "egg beater" blades of the common Darrieus design are replaced with straight vertical blade sections attached to the central tower with horizontal supports. This design is used by Shanghai based MUCE.

## Cycloturbines

Another variation of the Giromill is the Cycloturbine, in which each blade is mounted so that it can rotate around its own vertical axis. This allows the blades to be "pitched" so that they always have some angle of attack relative to the wind. The main advantage to this design is that the torque generated remains almost constant over a fairly wide angle, so a Cycloturbine with three or four blades has a fairly constant torque. Over this range of angles, the torque itself is near the maximum possible, meaning that the system also generates more power. The Cycloturbine also has the advantage of being able to self-start, by pitching the "downwind moving" blade flat to the wind to generate drag and start the turbine spinning at a low speed. On the downside, the blade pitching mechanism is complex and generally heavy, and some sort of wind-direction sensor needs to be added in order to pitch the blades properly.

## Helical Blades

The blades of a Darrieus turbine can be canted into a helix, e.g. three blades and a helical twist of 60 degrees. The original designer of the helical turbine is Ulrich Stampa. Gorlov proposed a similar design in 1995 (Gorlov's water turbines). Since the wind pulls each blade around on both the windward and leeward sides of the turbine, this feature spreads the torque evenly over the entire revolution, thus preventing destructive pulsations. This design is used by the Turby, Urban Green Energy, Enessere, Aerotecture and Quiet Revolution brands of wind turbine.

## Active Lift Turbine

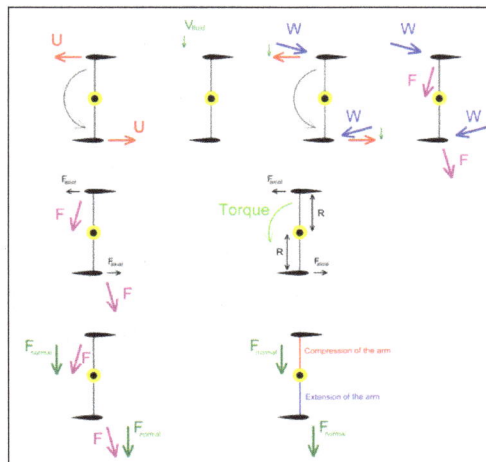

Active lift turbine - Axial and normal force.

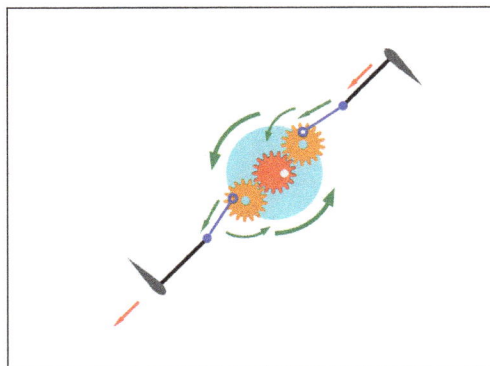

Active lift turbine - Crank rod system.

The relative speed creates a force on the blade. This force can be decomposed into an axial and normal force. In the case of a Darrieus turbine, the axial force associated with the radius creates a torque and the normal force creates on the arm a stress alternately for each half turn, a compression stress and an extension stress. With a crank rod system, the principle of the Active Lift Turbine is to transform this alternative constraint into an additional energy recovery.

## Savonius Wind Turbine

Savonius wind turbine.

Savonius wind turbines are a type of vertical-axis wind turbine (VAWT), used for converting the force of the wind into torque on a rotating shaft. The turbine consists of a number of aerofoils, usually—but not always—vertically mounted on a rotating shaft or framework, either ground stationed or tethered in airborne systems.

The Savonius wind turbine was invented by the Finnish engineer Sigurd Johannes Savonius in 1922. However, Europeans had been experimenting with curved blades on vertical wind turbines for many decades before this. The earliest mention is by the Italian Bishop of Czanad, Fausto Veranzio, who was also an engineer. He wrote in his 1616 book Machinae novae about several vertical axis wind turbines with curved or V-shaped blades. None of his or any other earlier examples reached the state of development made by Savonius. In his Finnish biography there is mention of his intention to develop a turbine-type similar to the Flettner-type, but autorotationary. He experimented with his rotor on small rowing vessels on lakes in his country. No results of his particular investigations are known, but the Magnus effect is confirmed by König. The two Savonius patents: US1697574, were filed in 1925 by Sigurd Johannes Savonius, and US1766765, in 1928.

### Operation

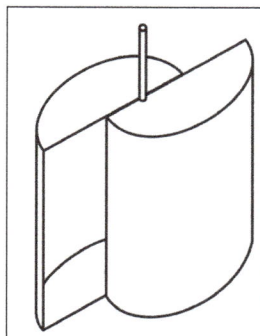

Schematic drawing of a two-scoop Savonius turbine.

The Savonius turbine is one of the simplest turbines. Aerodynamically, it is a drag-type device, consisting of two or three scoops. Looking down on the rotor from above, a two-scoop machine would look like an "S" shape in cross section. Because of the curvature, the scoops experience less drag when moving against the wind than when moving with the wind. The differential drag causes the Savonius turbine to spin. Because they are drag-type devices, Savonius turbines extract much less of the wind's power than other similarly-sized lift-type turbines. Much of the swept area of a Savonius rotor may be near the ground, if it has a small mount without an extended post, making the overall energy extraction less effective due to the lower wind speeds found at lower heights.

## Power and Rotational Speed

According to Betz's law, the maximum power that is possible to extract from a rotor is $P_{max} = \dfrac{16}{27}\rho \cdot r \cdot h \cdot v^3$, where $\rho$ is the density of air, $h$ and $r$ are the height and radius of the rotor and $v$ is the wind speed. However, in practice the extractable power is about half that (one can argue that only one half of the rotor — the scoop co-moving with the wind — works at each instant of time). Thus, one gets $P_{max} \approx 0.36 \mathrm{kgm^{-3}} \cdot h \cdot r \cdot v^3$.

The angular frequency of a rotor is given by $\omega = \dfrac{\lambda \cdot v}{r}$, where $\lambda$ is a dimensionless factor called the tip-speed ratio. λ is a characteristic of each specific windmill, and for a Savonius rotor λ is typically around unity.

For example, an oil-barrel sized Savonius rotor with h=1 m and r=0.5 m under a wind of v=10 m/s, will generate a maximum power of 180 W and an angular speed of 20 rad/s (190 revolutions per minute).

## Use

Savonius turbines are used whenever cost or reliability is much more important than efficiency.

Most anemometers are Savonius turbines for this reason, as efficiency is irrelevant to the application of measuring wind speed. Much larger Savonius turbines have been used to generate electric power on deep-water buoys, which need small amounts of power and get very little maintenance. Design is simplified because, unlike with horizontal axis wind turbines (HAWTs), no pointing mechanism is required to allow for shifting wind direction and the turbine is self-starting. Savonius and other vertical-axis machines are good at pumping water and other high torque, low rpm applications and are not usually connected to electric power grids. In the early 1980s, Risto Joutsiniemi developed a helical rotor version that does not require end plates, has a smoother torque profile and is self-starting in the same way a crossed pair of straight rotors.

The most ubiquitous application of the Savonius wind turbine is the Flettner rotor, which is commonly seen on the roofs of vans and buses and is used as a cooling device. The ventilator was developed by the German aircraft engineer Anton Flettner in the 1920s. It uses the Savonius wind turbine to drive an extractor fan. The vents are still manufactured in the UK by Flettner Ventilator Limited.

Small Savonius wind turbines are sometimes seen used as advertising signs where the rotation helps to draw attention to the item advertised. They sometimes feature a simple two-frame animation.

### Tethered Airborne Savonius Turbines

- Airborne wind turbines.

- Kite types.

- When the Savonius rotor axis is set horizontally and tethered, then kiting results. There are scores of patents and products that use the net lift Magnus effect that occurs in the autorotation of the Savonius rotor. The spin may be mined for some of its energy for making noise, heat, or electricity.

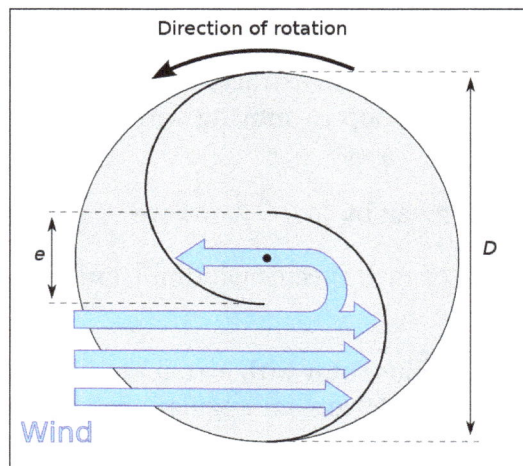

Operation of a Savonius turbine.

### References

- Windturbines: conserve-energy-future.com, Retrieved 19 February, 2019

- How-wind-turbine-works: energy.gov,  Retrieved 20 March, 2019

- Small-Wind-Turbines-Specification-Design-and-Economic-Evaluation: researchgate.net, Retrieved 21 April, 2019

- "Increase in the Savonius rotors efficiency via a parametric investigation". Researchgate. Retrieved 2017-06-02

- Practical-aspects-for-small-wind-turbine-applications: researchgate.net, Retrieved 22 May, 2019

# Wind Energy Conversion Systems

**3**

- **Principles of Wind Energy Conversion**

- **Classification of WECS**

- **Components of WECS**

- **Variable Speed Systems**

- **Grid Connected Systems**

Wind energy conversion system is used to convert the energy of wind movement into mechanical power which can be used to power machinery and electrical generator. It includes variable speed systems and grid connected systems. This chapter discusses these wind energy conversion systems in detail.

The wind energy conversion system (WECS) includes wind turbines, generators, control system, interconnection apparatus. Wind Turbines are mainly classified into horizontal axis wind turbines (HAWT) and vertical axis wind turbines (VAWT). Modern wind turbines use HAWT with two or three blades and operate either downwind or upwind configuration. This HAWT can be designed for a constant speed application or for the variable speed operation. Among these two types variable speed wind turbine has high efficiency with reduced mechanical stress and less noise. Variable speed turbines produce more power than constant speed type, comparatively, but it needs sophisticated power converters, control equipments to provide fixed frequency and constant power factor.

The generators used for the wind energy conversion system mostly of either doubly fed induction generator (DFIG) or permanent magnet synchronous generator (PMSG) type. DFIG have windings on both stationary and rotating parts, where both windings transfer significant power between shaft and grid. In DFIG the converters have to process only about 25-30 percent of total generated power (rotor power connected to grid through converter) and the rest being fed to grid directly from stator. Whereas, converter used in PMSG has to process 100 percent power generated, where 100 percent refers to the standard WECS equipment with three stage gear box in DFIG. Majority of wind turbine manufacturers utilize DFIG for their WECS due to the advantage in terms of cost, weight and size. But the reliability associated with gearbox, the slip rings and brushes in DFIG is unsuitable for certain applications. PMSG does not need a gear box and hence, it has high efficiency with less maintenance. The PMSG drives achieve very high torque at low speeds with less noise and require no external excitation. In the present trend WECS with multibrid concept is interesting and offers the same advantage for large systems in future. Multibrid is a technology where generator, gearbox, main shaft and shaft bearing are all integrated within a common housing. This concept allows reduce in weight and size of generators combined with the gear box technology. The generators with multibrid concept become cheaper and more reliable than that of the standard one, but it loses its efficiency.

To achieve high efficient energy conversion on these drives different control strategies can be implemented like direct torque control (DTC), field oriented control (FOC). The FOC using PI controller has linear regulation and the tuning becomes easier. The wind turbine electrical and mechanical parts are mostly linear and modeling will be easier. The blade aerodynamics of the wind turbine is a nonlinear one and hence the overall system model will become nonlinear. The wind energy conversion system which will be modeled as shown in figure may not be optimal for extracting maximum energy from the resource and hence various optimization techniques are used to achieve the goal.

## WECS Modeling

The basic device in the wind energy conversion system is the wind turbine which transfers the kinetic energy into a mechanical energy. The wind turbine is connected to the electrical generator through a coupling device gear train. The output of the generator is given to the electrical grid by employing a proper controller to avoid the disturbances and to protect the system or network.

shows the overall block diagram of the wind energy conversion system (WECS). Here, $V_\omega$ represents wind speed, $P_\omega$, $P_m$ and $P_e$ represent wind power, mechanical power and electrical power respectively.

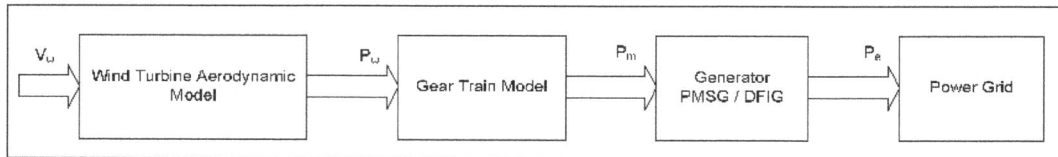

WECS Block Diagram.

## Wind Turbine

Wind energy is transformed into mechanical power through wind turbine and hence it is converted into electrical power. The mechanical power is calculated by using the following equation.

$$P_m = 0.5 A C_p (\lambda, B) v_{wind}$$

where, $\rho$ is the air density which normally takes the value in the range 1.22- 1.3 kg/m³, A is the area swept out by turbine blades (m²), $v_{wind}$ is the wind speed (m/s), $C_p(\lambda,\beta)$ is the power coefficient which depends on two factors: $\beta$, the blade pitch angle and the tip speed ratio and, $\lambda$, which is defined as:

$$\lambda = \Omega.R / v_{wind}$$

where, $\Omega$ is the angular speed (m/s) and R is the blade radius (m).

The power coefficient, $C_p$ is defined as:

$$C_p(\lambda,\beta) = C_1 \left( \frac{C_2}{\lambda_i} - C_3\beta - C_4 \right) \exp\left( \frac{-C_5}{\lambda_1} \right) + C_6\lambda$$

Where,

$$\frac{1}{\lambda_i} = \left( \frac{1}{\lambda + 0.08\beta} - \frac{0.035}{\beta^3 + 1} \right)$$

and coefficients $C_1 = 0.5176$, $C_2 = 116$, $C_3 = 0.4$, $C_4 = 5$, $C_5 = 21$, and $C_6 = 0.0068$.

The power coefficient is nonlinear, and it depends upon turbine blade aerodynamics and it can be represented as a function of tip speed ratio, $\lambda$. The optimum value of $\lambda$ corresponds to maximum of $C_p$ from the power coefficient-tip speed ratio curve.

Figure shows the power coefficient with respect to tip speed ratio. It is observed that the maximum power coefficient value $C_{p\_max}(\lambda,\beta) = 0.48$ for $\lambda = 12$ and for $\beta = 0°$. This particular value of $\lambda_{opt}$ results in optimal efficiency point where maximum power is captured from wind by the turbine.

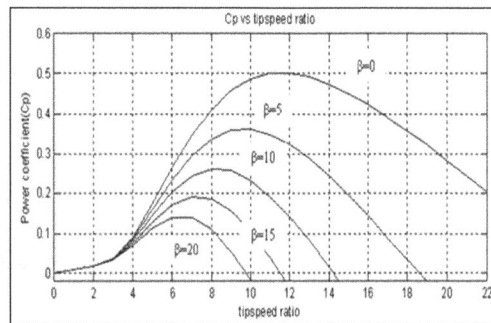

Power Coefficient of the Wind Turbine Model.

As the power transmitted is assumed to be the product of rotational speed with mechanical torque, the rotational torque is obtained as:

$$T_m = P_m / \Omega$$

Thus the optimal angular speed is achieved through the relation,

$$\Omega_{opt} = \lambda_{opt} v_{wind} / R$$

and the maximum mechanical power is,

$$P_{m-max} = 0.5\rho C_{p\max} v^3_{wind}$$

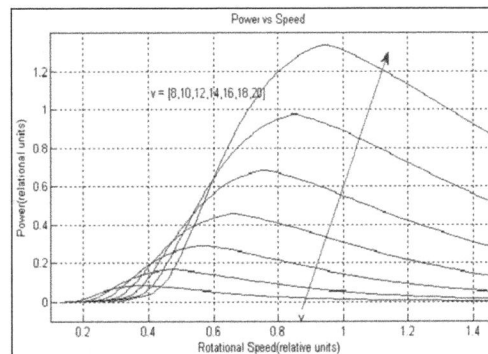

Wind Turbine Power Characteristics.

Figure. shows the wind turbine power characteristics obtained for various values of the wind tangential speed. Here it can be observed that maximum power (active) is achieved through optimal wind speeds and not at high wind velocity. The wind turbine does not operate when the wind speed is less than the minimum speed because the captured wind energy is not enough to compensate the losses and operation cost.

## Drive Train Modeling

The gear box enables to convert the slow speed of the turbine blades to high speed required by the electrical machine through the gear box ratio, $g_r$. This element converts the mechanical torque and the machine speed as follows:

$$T_{aero} = g_r . T_m$$
$$\Omega_m = g_r . \Omega$$

where, $T_{aero}$ is the aerodynamic torque and $\Omega_m$ is the machine speed. The drive train is modeled by the simplified motion equation as follows:

$$J \frac{d\Omega_m}{dt} = T_m - T_e f \Omega_m$$

where J is the mechanical inertia of the wind turbine and generator, $T_e$ is the electromagnetic torque, and f is the friction coefficient.

## PMSG Modeling

The modeling of PMSG type electrical equipment is made through the following equations, represented by d-q reference frame:

$$V_d = R_s i_d + L_d \frac{di_d}{dt} - \omega_e L_q i_q$$
$$V_q = R_s i_q + L_q \frac{di_q}{dt} - \omega_e L_d i_d + \omega_e \varphi_m$$

where $V_d$ and $V_q$ are d and q components of stator voltages (V), $i_d$ and $i_q$ are d and q components of stator currents (A), $R_s$ is stator resistance (ohms), $L_d$ and $L_q$ are machine inductances (H), $\omega_e$ is the electrical speed (rad/s) and $\varphi_m$ is the magnetic flux ($w_b$). The electrical torque is obtained through the following equation:

$$T_e = \frac{3}{2} p \left\{ \varphi_m i_q + \left( L_d - L_q \right) i_d i_q \right\}$$

where p is the pair of poles. The rotor dynamics of the machine is given by:

$$T_m - T_e = B \omega_r + J \frac{d\omega_r}{dt}$$

where B is the rotor friction (kgm² /s), J is the rotor inertia (kgm² ), $\omega_r$ is rotor speed (rad/s) and $T_m$ is the mechanical torque produced by wind (N$_m$). The machine dynamics can be simplified by assuming ($L_d = L_q = 0$) and the dreference current is zero ($i_d^* = 0$) and hence the product term ($L_d - L_q)i_d i_q$ is negligible. More information about the PMSG modeling is presented in.

## DFIG Modeling

The commonly used model for induction generator is the Park model. To obtain its state model we neglect stator resistance, and then we have:

$$\Phi_{sq} = 0, \frac{d\Phi_{sq}}{dt} = 0, V_{sd} = 0 \text{ and } V_{sq} = -V_s$$

where $V_s$ is the stator voltage. The reduced state model is obtained as follows:

$$V_{rd} = \sigma L_r \frac{di_{rd}}{dt} + r_r i_{rd} - (\omega_s - \Omega)\sigma L_r i_{rd}$$

$$V_{rd} = \sigma L_r \frac{di_{rd}}{dt} + r_r i_{rd} - (\omega_s - \Omega)\sigma L_r i_{rd} + \frac{M}{L}\Phi_{sq}(\omega_s - \Omega)$$

where $V_{rd}$ and $V_{rq}$ are the d and q components of rotor voltage (V), $V_{sd}$ and $V_{sq}$ are the d and q components of stator voltage (V), $i_{rd}$ and $i_{rq}$ are the d and q components of rotor current (A), $\Phi_{sd}$ and $\Phi_{sq}$ are the d and q components of stator flux ($w_b$), $\omega_s$ is the synchronous pulsation, $\Omega$ is the generator speed, $r_r$ is the rotor resistance, $L_s$ and $L_r$ are the stator and rotor inductances (H), M is the mutual inductance (H) and $\sigma$ is the leakage parameter given as, $\sigma = 1 - (M^2/L_sL_r)$. More information about stand-alone and grid connected operation of DFIG is presented in papers.

DFIG is one of the most important generators and are widely used for variable speed WECS. Nowadays, this type of DFIG-WECS has a part of wind energy market, which is close to 50%. Major wind turbine producers manufacture wind energy conversion system based on DFIG's but the difficulties associated in complying with grid side ride-through requirement may limit its use in future. To overcome the above stated difficulty, PMSG technology looks most promising in WECS, which shares nearly 45% in wind energy market. Multi-pole PMSG's with full power back-to-back converters appears to be the configuration to be adapted by most of the wind turbine manufacturers in near future, gradually replacing DFIG in wind energy market. Wind energy conversion systems with the range of 1.5 MW to 3 MW are manufactured by 44 companies in 163 models. Among these 60 models uses DFIG, 66 models uses PMSG, 18 models uses cage type IG and 19 models uses synchronous generator with external excitation.

## Control Strategies

## DFIG-WECS Control

Controller is designed to adjust the turbine speed to extract the maximum power from wind source. Usually PI controller is designed for this purpose according to the estimated parameters either offline or online and the observed disturbance torque is feed forward to increase system robustness. But this type of classical controller is not enough to serve the purpose efficiently. Hence various improvements are made for the controller to achieve the requirement as stated below.

The general structure of control block diagram in the DFIG-WECS having two levels of control. The lower level control being the electrical control system, i.e. torque and reactive power control. The rotor side controller (RSC) and grid side controller (GSC) are employed to serve the electrical control system. To compromise the dynamics of variable speed WECS Poitiers has implemented a two degree freedom Regulation-Solution-Tracking (R-S-T) controller, which is a widely preferred digital controller and Linear Quadratic Gaussian (LQG) controller with state feedback. For the ease of implementation of controller tuning the LQG controller is preferred than R-S-T controller.

To reduce the output power fluctuations of DFIG with a flywheel energy storage system (FESS) Jerbi has proposed a fuzzy supervisor using sugeno fuzzy model based on two parameters flywheel speed and wind power accessibility so as to ensure a smooth reactive power to the load supplied by the wind generator. This ensures to reduce the voltage fluctuations. For the variable speed WECS has proposed a method which will combine fuzzy neural network and sliding mode speed observer. This technique allows faster convergence to a simple linear dynamic behavior, even in presence of parameter changes and uncertainties which are the major problems expected from variable speed drive control. As most of the controllers depends on torque control M. Pucci proposes a method based on speed control of the machine. This is achieved through total least squares (TLS) EXIN full order observer which turns as an intelligent sensor less technique for wind generation control.

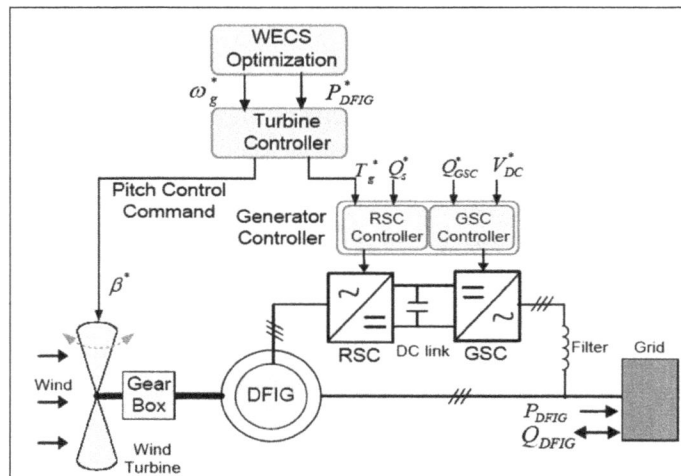

Control Block Diagram of DFIG Based WECS.

There are other types of controllers using PID for controlling the WECS with optimal power output. Muhando et.al proposes a self-tuning regulator by incorporating recursive least square algorithm to predict the process parameters and update the states. This regulator gives more efficient compared to classical PID controller. To achieve optimal power, adaptive fuzzy PID control strategies are proposed and proved the performance is improved than that of conventional one.

Many research papers discuss the intelligent controllers with sensor less wind energy control. Whei-Min Lin proposed a technique where design is made online through recurrent fuzzy neural controller (RFNN) with high performance model reference adaptive system (MRAS) observer which replaces the use of sensors to measure the parameters. Lin et.al suggested that to increase the learning ability of back propagation network of RFNN modified particle swarm optimization is used (mPSO). This provides fast and accurate velocity information to avoid anemometers and rotor position is estimated from flux linkages.

Behra and Rao investigated novel grid side controller, modeled in synchronously rotating reference frame to improve the dynamic response on various load conditions. Liu and Hsu discuss the stator resistance effect on the excitation voltage, which gives the maximum output power and minimal loss for the DFIG. The DFIG-WECS model can be validated through various simulation tools. To improve the real time simulation time FPGA's can be incorporated using Runge-Kutta numerical integration algorithm in compared to PC based simulation.

## PMSG-WECS Control

Figure shows the block diagram of PMSG based WECS with two stages as optimization and electrical controllers. The various techniques are discussed as below.

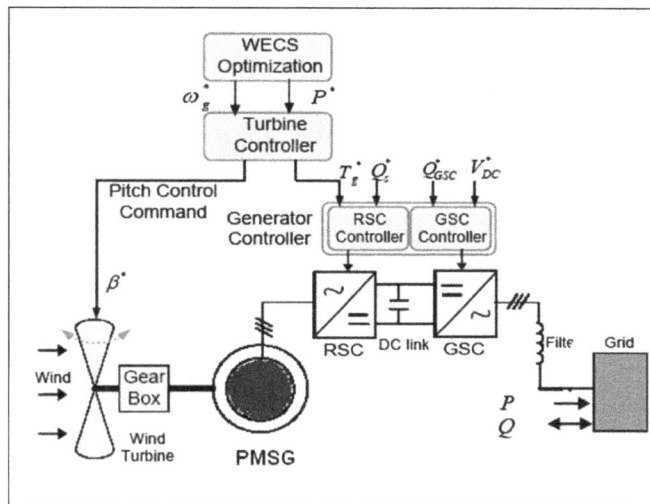

Control Block Diagram of PMSG Based WECS.

Jordi Zaragoza discusses the control of PMSG based WECS using field oriented control (FOC) for controlling speed regulation and generating current. A PI controller is used for this purpose and the tuning parameters are determined through zero-pole cancellation (ZPC) strategy. Jemaa Brahmi et.al compares three control methods for PMSG based WECS. In the first model MRAS observer is used to estimate the parameters of the controller by comparing output of reference model and adaptive model. Secondly artificial neural network (ANN) based observer were trained to produce desired output correction to the estimated speed. And lastly, sliding mode observer (SMO) based control is discussed. Among these three techniques SMO and ANN observer gives good results when static resistance varies and chattering phenomenon is observed with SMO. The static resistance causes a static error in case of MRAS observer. The SMO is more robust than the other two methods.

Kelvin Tan and Syed Islam proposed a sensor less optimal control strategy using fuzzy power mapping technique. Using the result of power mapping loop and alternator frequency derivative loop, the controller allows the bus voltage to vary to maximum power operating point. Bao proposes an adaptive control strategy for variable speed wind turbine using neural based identifier and inverse model controller which has good learning capability and high performance than nonlinear controller. Wang et.al proposes indirect matrix converter of PID type which has more degree of freedom and can be implemented in an easy way for the PMSG based WECS. This shows excellent dynamic performances with lower mechanical stress and robust in control.

There are other techniques for controlling the WECS. Wind prediction control scheme is the one where the control parameters are obtained through forecasting the wind using autoregressive statistical models. This method will avoid the mechanical sensors used to get the actual parameter tracking.

## Optimization of WECS

The next challenging task in WECS is to operate the system at the optimal point so as to increase the output. The optimization block is shown in Figs. and for DFIG and PMSG based WECS. Among the various techniques followed for this, maximum power point tracking (MPPT) plays a vital role.

Kesraoui proposed the MPPT algorithm using step and search method to extract the maximum power from the wind turbine from cut-in to rated wind velocity by sensing the dc link power. Iulian Munteanu used wind turbulence as a search for MPPT instead of sinusoidal search. Here high speed shaft's average speed is slowly adjusted using FFT of measures from system as an estimate to the optimal point. This method is effective when the wind turbulence level is high. The same author also proposes sliding mode control, optimal regime tracking and frequency separation principle for optimizing the WECS based on PMSG type. Syed Muhammad et.al proposes a novel method of MPPT along with Hill Climb Search (HCS) algorithm. This method performs self-tuning to cope with the non-constant efficiency of the generator- converter systems. This paper ensures high control efficiency and faster tracking under rapid wind change without requiring any prior knowledge of system characteristics.

Adaptive neuro fuzzy inference system (ANFIS) based MPPT is discussed in Meharrar. Here variation of wind speed is used as input for ANFIS to predict optimal speed rotation and back propagation learning algorithm is used for training this network. This method shows better response than compared to FLC. Other type of fuzzy controller was proposed by Calderaro using Takagi-Sugeno-Kang (TSK) model to maximize energy extraction which avoids sensors in MPPT. To estimate the optimum value for generator support vector regression (SVR) algorithm is used by Lee. It is proved that the generator losses are reduced up to 40% and also excellent in accuracy. Kazmi proposes growing neural gas (GNG) based MPPT which is having FOC integrated with TLS EXIN observer without sensors to optimize the system.

A new method of MPPT which uses gradient approximation algorithm is proposed by Ying-Yi Hong. This method does not require anemometer, tachometer and information of wind turbine generator system characteristics. As the gradient approximation converges very fast and the implementation is easy, this method is efficient for an optimization problem. Kongnam and Nuchprayoon propose to apply PSO to solve for optimum rotor speed in fixed and variable speed operation of WECS. In this power and energy are more dependent of mean wind speed than weibull distribution of speed.

Pitch angle control of wind turbine turns out to be another optimization problem in WECS. Yilmaz and Zafer discusses radial basis function based neural network controller which tracks the reference signal based on training to achieve better controller. Boukhezzar propose control strategy by combining torque control and pitch control which leads to good performance in rotor speed regulation and electrical power regulation with acceptable control loads compared to classical PID and LQG controllers.

## Principles of Wind Energy Conversion

The power in the wind is proportional to the wind speed cubed; the general formula for power in the wind is:

$$P = \frac{1}{2}\rho A V^3$$

where P is the power available in watts, p is the density of air (which is approximately 1.2kg/m³ at sea level), A is the cross-section (or swept area of a windmill rotor) of air flow of interest and V is the instantaneous free-stream wind velocity. If the velocity, V, is in m/s (note that lm/s is almost exactly 2 knots or nautical miles per hour), the power in the wind at sea level is:

$$P = 0.6 V^3 \text{ watts/ m}^2 \text{ of rotor area}$$

Because of this cubic relationship, the power availability is extremely sensitive to wind speed; doubling the wind speed increases the power availability by a factor of eight; Table indicates this variability.

Table: Power in the wind as a function of wind speed in units of power per unit area of wind stream.

| wind speed | m/s | 2.5 | 5 | 7.5 | 10 | 15 | 20 | 30 | 40 |
|---|---|---|---|---|---|---|---|---|---|
| | km/h | 9 | 18 | 27 | 36 | 54 | 72 | 108 | 144 |
| | mph | 6 | 11 | 17 | 22 | 34 | 45 | 67 | 90 |
| power density | kW/m² | .01 | .08 | .27 | .64 | 2.2 | 5.1 | 17 | 41 |
| | hp/ft² | .001 | .009 | .035 | .076 | .23 | .65 | 2.1 | 5.2 |

This indicates the very high variability of wind power, from around 10W/m2 in a light breeze up to 41 000Wm² in a hurricane blowing at 144km/h. This extreme variability greatly influences virtually all aspects of system design. It makes it impossible to consider trying to use winds of less than about 2.5m/s since the power available is too diffuse, while it becomes essential to shed power and even shut a windmill down if the wind speed exceeds about 10-15m/s (25-30mph) as excessive power then becomes available which would damage the average windmill if it operated under such conditions.

The power in the wind is a function of the air-density, so it declines with altitude as the air thins, as indicated in table.

Table: Variation of air density with altitude.

| altitude (ft) | 0 | 2 500 | 5 000 | 7 500 | 10 000 |
|---|---|---|---|---|---|
| a.s.l. (m) | 0 | 760 | 1 520 | 2 290 | 3 050 |
| density correction factor | 1.00 | 0.91 | 0.83 | 0.76 | 0.69 |

Because the power in the wind is so much more sensitive to velocity rather than to air density, the effect of altitude is relatively small. For example the power density of a 5m/s wind at sea level is about 75 watts/m²; however, due to the cube law, it only needs a wind speed of 5.64m/s at 3 000m a.s.l. to obtain exactly the same power of 75 watts/m². Therefore the drop in density can be compensated for by quite a marginal increase in wind velocity at high altitudes.

## Energy Available in the Wind

Because the speed of the wind constantly fluctuates, its power also varies to a proportionately greater extent because of the cube law. The energy available is the summed total of the power over a given time period. This is a complex subject (Lysen gives a good introduction to it). The usual starting point to estimate the energy available in the wind at a specific location is some knowledge of the mean or average wind speed over some predefined time period; typically monthly means may be used. The most important point of general interest is that the actual energy available from the wind during a certain period is considerably more than if you take the energy that would be produced if the wind blew at its mean speed without variation for the same period.

Typically the energy available will be about double the value obtained simply by multiplying the instantaneous power in the wind that would correspond to the mean wind speed blowing continuously, by the time interval. This is because the fluctuations in wind speed result in the average power being about double that which occurs instantaneously at the mean wind speed. The actual factor by which the average power exceeds the instantaneous power corresponding to the mean windspeed can vary from around 1.5 to 3 and depends on the local wind regime's actual variability. The greater the variability the greater this factor.

However, for any specific wind regime, the energy available will still generally be proportional to the mean wind speed cubed.

## Converting Wind Power to Shaft Power

There are two main mechanisms for converting the kinetic energy of the wind into mechanical work; both depend on slowing the wind and thereby extracting kinetic energy. The crudest, and least efficient technique is to use drag; drag is developed simply by obstructing the wind and creating turbulence and the drag force acts in the same direction as the wind. Some of the earliest and crudest types of wind machine, known generically as "panamones", depend on exposing a flat area on one side of a rotor to the wind while shielding (or reefing the sails) on the other side; the resulting differential drag force turns the rotor.

The other method, used for all the more efficient types of windmill, is to produce lift. Lift is produced when a sail or a flat surface is mounted at a small angle to the wind; this slightly deflects the wind and produces a large force perpendicular to the direction of the wind with a much smaller drag force. It is this principle by which a sailing ship can tack at speeds greater than the wind. Lift mainly deflects the wind and extracts kinetic energy with little turbulence, so it is therefore a more efficient method of extracting energy from the wind than drag.

It should be noted that the theoretical maximum fraction of the kinetic energy in the wind that could be utilized by a "perfect" wind turbine is approximately 60%. This is because it is impossible to stop the wind completely, which limits the percentage of kinetic energy that can be extracted.

## Horizontal and Vertical Axis Rotors

Windmills rotate about either a vertical or a horizontal axis. All the windmills illustrated so far, and most in practical use today, are horizontal axis, but research is in progress to develop vertical axis machines. These have the advantage that they do not need to be orientated to face the wind, since they present the same cross section to the wind from any direction; however this is also a disadvantage as under storm conditions you cannot turn a vertical axis rotor away from the wind to reduce the wind loadings on it.

There are three main types of vertical axis windmill. Panamone differential drag devices (mentioned earlier), the Savonius rotor or "S" rotor and the Darrieus wind turbine. The Savonius rotor consists of two or sometimes three curved interlocking plates grouped around a central shaft between two end caps; it works by a mixture of differential drag and lift. The Savonius rotor has been promoted as a device that can be readily improvised on a self-build basis, but its apparent simplicity is more perceived than real as there are serious problems in mounting the inevitably heavy rotor securely in bearings and in coupling its vertical drive shaft to a positive displacement pump (it turns too slowly to be useful for a centrifugal pump). However the main disadvantages of the Savonius rotor are two-fold:

- It is inefficient, and involves a lot of construction material relative to its size, so it is less cost-effective as a rotor than most other types.

- It is difficult to protect it from over-speeding in a storm and flying to pieces.

The Darrieus wind turbine has airfoil cross-section blades (streamlined lifting surfaces like the wings of an aircraft). These could be straight, giving the machine an "H"-shaped profile, but in practice most machines have the curved "egg-beater" or troposkien profile as illustrated. The main reason for this shape is because the centrifugal force caused by rotation would tend to bend straight blades, but the skipping rope or troposkien shape taken up by the curved blades can resist the bending forces effectively. Darrieus-type vertical axis turbines are quite efficient, since they depend purely on lift forces produced as the blades cross the wind (they travel at 3 to 5 times the speed of the wind, so that the wind meets the blade at a shallow enough angle to produce lift rather than drag).

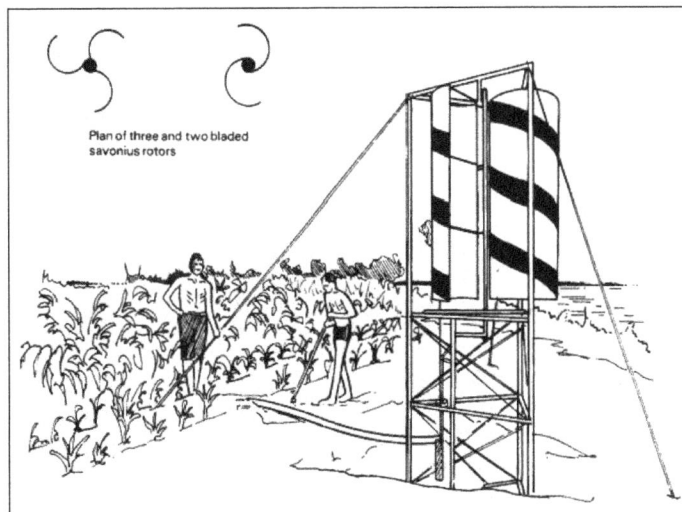

Savonius Rotor vertical-axis windpump in Ethiopia. It was found to be less cost- effective than the 'Cretan' windmill.

The Darrieus was predated by a much cruder vertical axis windmill with Bermuda (triangular) rig sails from the Turks and Caicos Islands of the West Indies. This helps to show the principle by which the Darrieus works, because it is easy to imagine the sails of a Bermuda rig producing a propelling force as they cut across the wind in the same way as a sailing yacht; the Darrieus works on exactly the same principle.

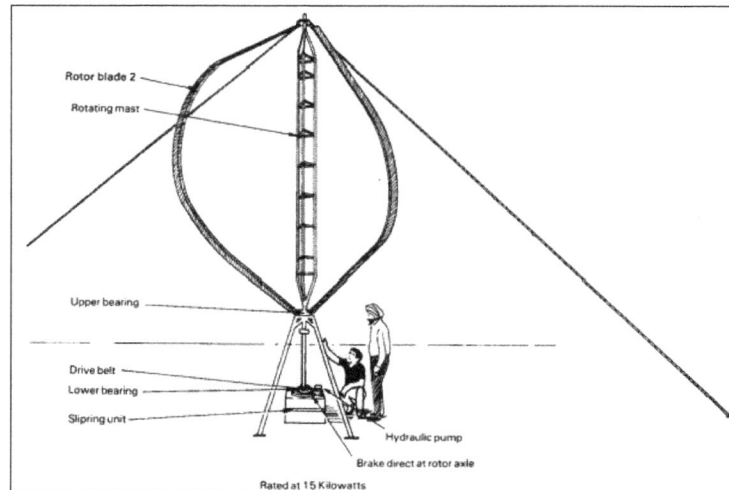

Typical Troposkien shaped Darrieus vertical axis wind turbine.

There are also two main types of Darrieus wind turbine which have straight blades; both control overspeed and consequent damage to the blades by incorporating a mechanism which reefs the blades at high speeds. These are the Variable Geometry Vertical Axis Wind Turbine (VGVAWT) developed by Musgrove in the UK and the Gyromill Variable Pitch Vertical Axis Wind Turbine (VPVAWT), developed by Pinson in the USA. Although the Musgrove VGVAWT has been tried as a windpump by P I Engineering, all the current development effort is being channelled into developing medium to large electricity grid-feeding, vertical-axis wind-generators, of little relevance to irrigation pumping.

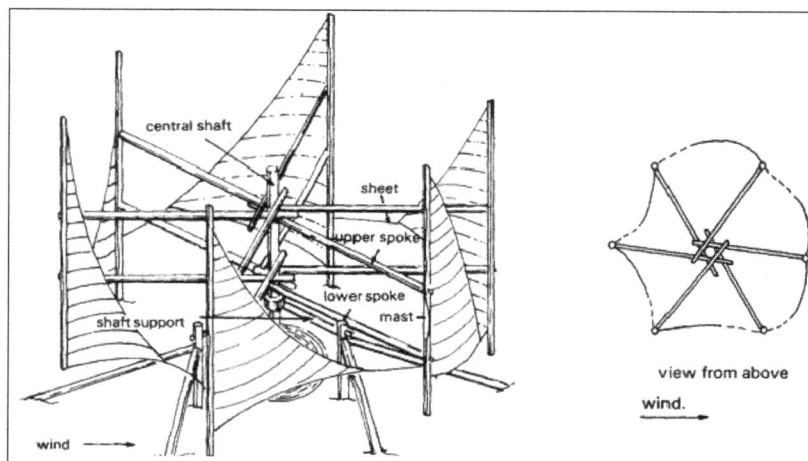

Turks and Caicos islands vertical-axis sail rotor.

Vertical axis windmills are rarely applied for practical purposes, although they are a popular subject for research. The main justification given for developing them is that they have some prospect of being simpler than horizontal axis windmills and therefore they may become more cost-effective. This still remains to be proved.

Most horizontal axis rotors work by lift forces generated when "propeller" or airscrew like blades are set at such an angle that at their optimum speed of rotation they make a small angle with the wind and generate lift forces in a tangential direction. Because the rotor tips travel faster than the roots, they "feel" the wind at a shallower angle and therefore an efficient horizontal axis rotor requires the blades to be twisted so that the angle with which they meet the wind is constant from root to tip. The blades or sails of slow speed machines can be quite crude but for higher speed machines they must be accurately shaped airfoils; but in all three examples illustrated, the principle of operation is identical.

## Efficiency, Power and Torque Characteristics

Any wind turbine or windmill rotor can be characterized by plotting experimentally derived curves of power against rotational speed at various windspeeds. Similarly the torque produced by a wind rotor produces a set of curves.

The maximum efficiency coincides with the maximum power output in a given windspeed. Efficiency is usually presented as a non-dimensional ratio of shaft-power divided by wind-power passing through a disc or shape having the same area as the vertical profile of the windmill rotor; this ratio is known as the "Power Coefficient" or $C_p$ and is numerically expressed as:

$$C_p = \frac{T}{\frac{1}{2}\rho A V^3}$$

The speed is also conventionally expressed non-dimensionally as the "tip-speed ratio" ( /?/. ). This is the ratio of the speed of the windmill rotor tip, at radius R when rotating at $\omega$ radians/second, to the speed of the wind, V, and is numerically:

$$\lambda = \frac{\omega R}{V}$$

When the windmill rotor is stationery, its tip-speed ratio is also zero, and the rotor is stalled. This occurs when the torque produced by the wind is below the level needed to overcome the resistance of the load. A tip-speed ratio of 1 means the blade tips are moving at the same speed as the wind (so the wind angle "seen" by the blades will be 45°) and when it is 2, the tips are moving at twice the speed of the wind, and so on.

The $C_p$ versus curves for three different types of rotor, with configurations A, B, C, D, El, E2 and F as indicated, are shown in figure The second set of curves show the torque coefficients, which are a non-dimensional measure of the torque produced by a given size of rotor in a given wind speed (torque is the twisting force on the drive shaft). The torque coefficient, $C_t$, is defined as:

$$C_t = \frac{T}{\frac{1}{2}\rho A V^2 R}$$

where T is the actual torque at windspeed V for a rotor of that configuration and radius R.

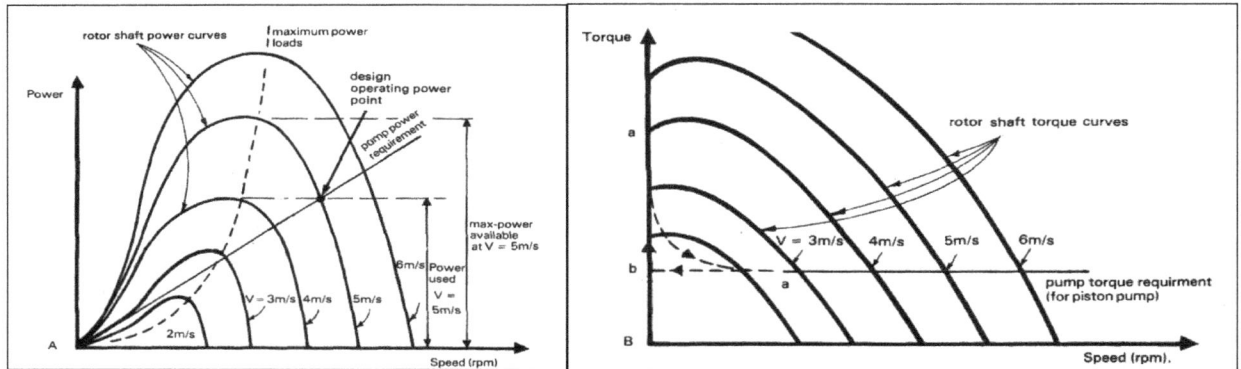

The power (A) and torque (B) of a wind rotor as a function of rotational speed for difference wind speeds.

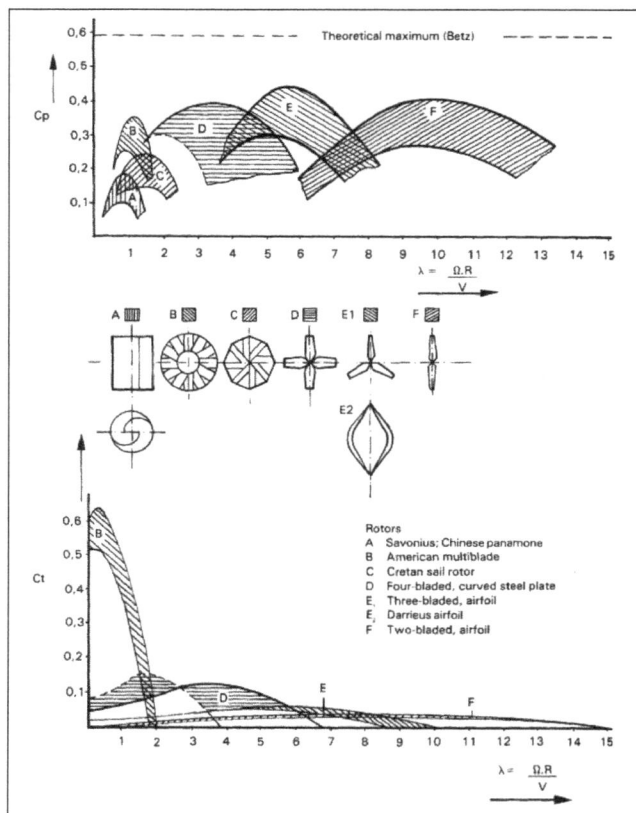

The power coefficients ($C_p$) (above) and the torque coefficients ($C_t$) of various types of wind turbine rotor plotted against tip-speed ratio ($\lambda$).

## Rotor Solidity

"Solidity" (ó) is a fairly graphic term for the proportion of a windmill rotor's swept area that is filled with solid blades. It is generally defined as the ratio of the sum of the width, or "chords" of all the blades to the circumference of the rotor; i.e. 24 blades with a chord length (leading edge to trailing edge) of 0.3m on a 6m diameter rotor would have a tip solidity of:

$$\sigma = \frac{(24 \times 0.3)}{6\pi}$$

Multi-bladed rotors, as used on windpumps, are said to have high "solidity", because a large proportion of the rotor swept area is "solid" with blades. Such machines have to run at relatively low speeds and will therefore have their blades set at quite a coarse angle to the plain of rotation, like a screw with a coarse thread. This gives it a low tip-speed ratio at its maximum efficiency, of around 1.25, and a slightly lower maximum coefficient of performance than the faster types of rotor such as "D", "E" and "F" in the figure. However, the multi-bladed rotor has a very much higher torque coefficient at zero tip-speed ratio (between 0.5 and 0.6) than any of the other types. Its high starting torque (which is higher than its running torque) combined with its slow speed of rotation in a given wind make it well-suited to driving reciprocating borehole pumps.

In contrast, the two or three-bladed, low-solidity, rotors "El" and "F", are the most efficient, (with the highest values for $C_p$), but their tips must travel at six to ten times the speed of the wind to achieve their best efficiency. To do so they will be set at a slight angle to the plain of rotation, like a screw with a fine thread and will therefore spin much faster for a given windspeed and rotor diameter than a high solidity rotor. They also have very little starting torque, almost none at all, which means they can only start against loads which require little torque to start them, like electricity generators (or centrifugal pumps) rather than positive displacement pumps.

All this may sound academic, but it is fundamental to the design of wind rotors; it means that multi-bladed "high-solidity" rotors run at slow speeds and are somewhat less efficient than few-bladed "low solidity" rotors, but they have typically five to twenty times the starting torque.

## Matching Rotors to Pumps

High solidity rotors are typically used in conjunction with positive displacement (piston) pumps, because, as explained in section, single-acting piston pumps need about three times as much torque to start them as to keep them going. Low solidity rotors, on the other hand, are best for use with electricity generators or centrifugal pumps, or even ladder pumps and chain and washer pumps, where the torque needed by the pump for starting is less than that needed for running at design speed. Table indicates the relative characteristics and $C_p$ values for various typical wind rotor types so far described.

Table: Comparison between different rotor types.

| Type | Performance characteristic | Manufacturing requirements | Cp | Solidity 6 | t.s.r. (Optimum) (t.s.r. = tip-speed ratio (X)) |
|---|---|---|---|---|---|
| Horizontal axis — | | | | | |
| Cretan sail or flat paddles | Medium starting torque and low speed | Simple | 0.05 to 0.15 | 50% 1.5-2.0 | 1.5-2.0 |
| Cambered plate fan (American) | High starting torque and low speed | Moderate | 0.15 to 0.30 | 50 to 80% | 1-1.5 |
| Moderate speed aero-generator | Low starting torque and moderate speed | Moderate, but with some precision | 0.20 to 0.35 | 5 to 10% | 3-5 |
| High speed aero-gen. | Almost zero starting torque and high speeds | Precise | 0.30 to 0.45 | under 5% | 5-10 |

| Vertical axis — | | | | | |
|---|---|---|---|---|---|
| Panemone | Medium starting torque and low speed | Simple | under 0.10 | 50% | .4-.8 |
| Savonius rotor | Medium starting torque and moderate speed | Moderate | 0.15 | 100% | .8-1 |
| Darrieus rotor | Zero starting torque and moderate speed | Precise | 0.25 to 0.35 | 10% to 20% | 3-5 |
| VGVAWT or Gyromill | Zero or small starting torque and moderate speed | Precise | 0.20 to 0.35 | 15% to 40% | 2-3 |

The load lines for a positive displacement direct-driven pump superimposed on the wind rotor output curves. The dotted line on indicates the locus of the points of maximum power; the system will only function continuously when the operating point is to the right of the line of maximum power, as under that condition any slight drop in wind speed causes the machine to slow and the power absorbed by the shaft to increase, which results in stable operation. The operating point can only remain to the left of the maximum power locus under conditions of increasing windspeed. It can be seen that the positive displacement pump requires more or less constant torque of 10Nm in the example, once rotation has been established, but it needs over three times as much torque to start it for reasons explained in section. The torque curves in figure B indicate that 5m/s windspeed is needed to produce the torque required to start the windpump rotating, but once rotation has commenced, the windspeed can fall to 3m/s before the operating point moves to the left of th maximum powere locus and the windpump will stop. Note that the broken line a'-a represents a transient condition that only occurs momentarily when the windpump starts to rotate.

To extract the maximum power from a windpump at all times would require a load which causes the operating point to follow close to the locus of maximum power.The figure also indicates that that the operating point will always be where the windpump rotor curve for the windspeed prevailing at a given moment coincides with the pump load line. In the example, the operating point is shown for a windspeed of 5m/s; in this example, it can be seen that only about two-thirds of the maximum power that could be produced in this wind speed is used by the pump, because its load line diverges from the cubic maximum power curve. This discrepancy is a mis-match between the prime-mover (the windmill rotor) and the load (the pump). The proportion of the power available from the rotor in a given windspeed which is usefully applied is known as the "matching efficiency", and is analysed in detail in Pinilla. The figure illustrates how this mis-match becomes progressively worse as the wind speed increases. This mis-match is actually less serious than it may seem, since the time when the best efficiency is needed is at low windspeeds when, fortunately, the best efficiency is achieved. When a windmill is running fast enough to be badly matched with its pump, it means that the wind is blowing more strongly than usual and the chances are that the output, although theoretically reduced by bad matching, will be more than adequate, as the extra speed will compensate for the reduction in efficiency.

It may be thought that centrifugal pumps would match better with a windmill than positive displacement pumps, but in practice their efficiency falls rapidly to zero below a certain threshold running speed at a fixed static head. In otherwords, centrifugal pumps do not readily run with adequate efficiency over as wide a speed range as is necessary to match most windmills rotors and they are therefore not generally used with windmills (except with intermediate electrical transmission which can modify the relationship between the pump and windmill speeds).

When generators are used as a load, instead of pumps, a much better match can be obtained. Wind generators therefore tend to have a better matching efficiency over their whole range of operating speeds than windpumps; the interested reader is referred to a text on this subject, such as Lysen.

There is considerable scope for improving the overall performance of wind pumps by developing methods of improving the rotor-to-pump match over a wider range of windspeeds; a certain amount of, work is being carried out in this field and if successful could result in considerably more effective windpumps in the future. But in the meantime the main problem is to choose the most appropriate pump size for a given windmill in a given wind regime and location. figure shows how the pump load line can be altered simply by changing the mean pump rod pull, either by changing the stroke (by lengthening or shortening the crank) or by changing the diameter of the pump being used. A longer stroke and a larger pump will increase the pump rod force, and increase the mean torque requirement and hence the slope of the load line, and vice- versa. In figure it is clear that increasing the load increases the hydraulic output at higher speeds, but it also increases the value of $V_s$, the starting windspeed. Therefore, pump "C" in the diagram will start in a much lighter wind than the other pumps, but because of the shallower load line the output will be much smaller in high winds. There is therefore an important trade-off between achieving starting in adequately light winds and achieving a good output.

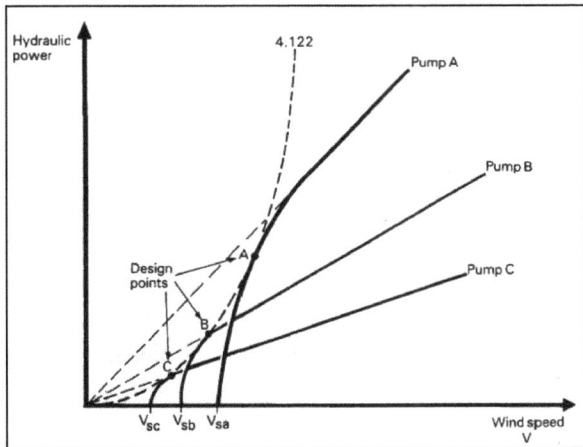

The trade-off between starting windspeed and output for differently loaded windpumps.

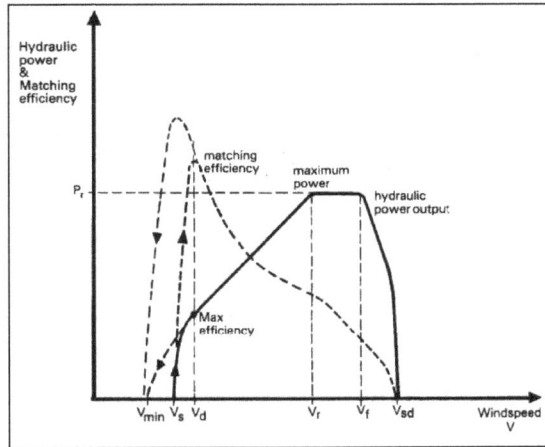

The operating characteristic of a windpump showing how the power output and matching efficiency vary with windspeed.

The operating characteristic of a typical windpump, given in Figure, shows how if the start-up windspeed is $V_s$ a windpump can run down to a slightly lower windspeed V (as explained earlier, assuming the use of a piston pump). It reaches its best match with the rotor at windspeeds close to $V_{min}$ (in theory at 0.8VS) which is the "Design Windspeed", and then increases its output almost linearly with windspeed to V (its rated windspeed). At still higher windspeeds means must be introduced to prevent it speeding up further, or the machine may be over-loaded and damaged or destroyed; various methods for doing this are discussed in the next section below. At very high windspeeds, the only safe course of action is to make the windmill "reef", "furl" or "shut- down";

the figure shows how this process commences at a windspeed $V_f$ (furling speed) and is completed at windspeed $v_{sd}$ (shut-down).

## Methods of Storm Protection and Furling

Windmills must have a means to limit the power they can deliver, or else they would have to be built excessively strongly (and expensively) merely to withstand only occasional high power outputs in storms. Sailing ships "take in canvas" by wholly or partially furling the sails (manually) when the wind is too strong, and Cretan sail windmills and other such simple traditional designs generally use exactly the same technique; fewer sails are used in high winds or else the sails are partially rolled around their spars. Metal farm windmills, however, have fixed steel blades, so the solution most generally adopted is to mount the rotor offset from the tower centre so that the wind constantly seeks to turn the rotor behind the tower. Under normal conditions the rotor is held into the wind by a long tail with a vane on it. This vane is hinged, and fixed in place with a pre-loaded spring (as illustrated), then when the wind load on the rotor reaches a level where the force is sufficient to overcome the pre-tension in the spring, the tail will start to fold until the wind pushes the rotor around so that it presents its edge to the wind, as in figure. This furling process starts when the rated output is reached and if the windspeed continues to rise, it increases progressively until the machine is fully furled. Then when the wind drops, the spring causes the tail vane to unfold again and turn the rotor once again to face the wind. On commercial farm windmills, this action is normally completely automatic.

Wind-generators and other windturbines with high speed, low-solidity rotors often use a mechanism which changes the blade pitch; e.g. the Dunlite machine of figure which has small counter-weights, visible near the rotor hub, which force the blades into a coarser pitch under the influence of centrifugal force when the rotor reaches its furling speed, against the force of a spring enclosed in the hub. Alternatively air-brake flaps are deployed to prevent overspeed. Larger windturbines do not use tail vanes to keep them facing the wind, as they cannot stand being yawed as fast as might occur if there is a sudden change in wind direction. Instead they usually have a worm-reduction gear mechanism similar to that in a crane, which inches them round to face the wind; this can be electrically powered on signals from a small wind direction vane, or it can use the mechanism visible on the Windmatic in figure, used on large windmills for several centuries, where a sideways mounted windrotor drives the orientating mechanism every time the main rotor is at an angle other than at right angles to the wind direction.

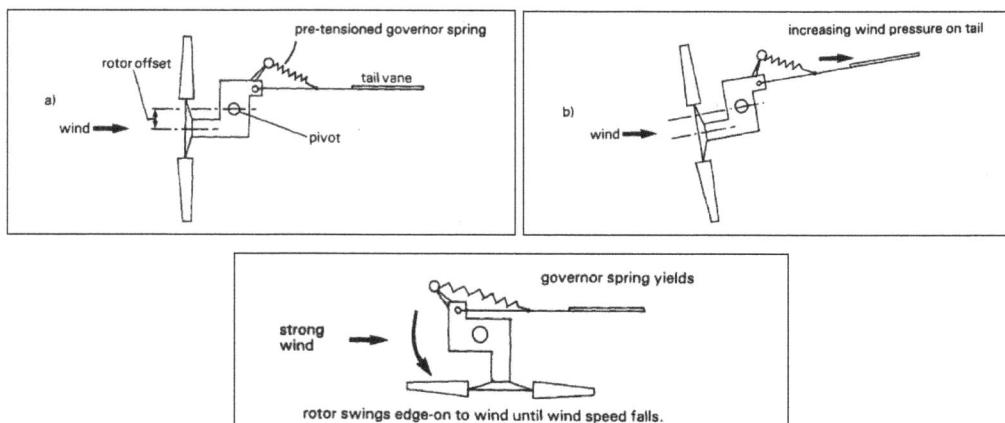

Typical windpump storm protection method in which rotor is yawed edge-on to the wind (plan view).

# Classification of WECS

- Based on axis:
    - Horizontal axis machines.
    - Vertical axis machines.
- According to size:
    - Small size machines (upto 2k W).
    - Medium size machines (2 to 100k W).
    - Large size machines (100k W and above).
        - Single generator at single site.
        - Multiple generators.
- Types of output:
    - DC output.
        - DC generator.
        - Alternator rectifier.
- AC output:
    - Variable frequency, variable or constant voltage AC.
    - Constant frequency, variable or constant voltage AC.
- According to the rotational speed of the area turbines:
    - Constant speed and variable pitch blades.
    - Nearly constant speed with fixed pitch blades.
    - Variable speed with fixed pitch blades.
        - Field modulated system.
        - Double output indication generator.
        - AC-DC-AC link.
        - AC commentator generator.
    - Variable speed constant frequency generating system.
- As per utilization of output:
    - Battery storage.
    - Direct conversion to an electro magnetic energy converter.

- ◦ Thermal potential.

- ◦ Inter convention with conventional electric utility guides.

## Components of WECS

Early wind machines ranged in their rated powers from 50 to 100 kW, with rotor diameters from 15 to 20 meters. Commercial wind turbines now have ratings over 1 MW and machines for the land based and offshore applications have rated power outputs reaching 5 and even 7-10 MW of rated power for off-shore wind applications.

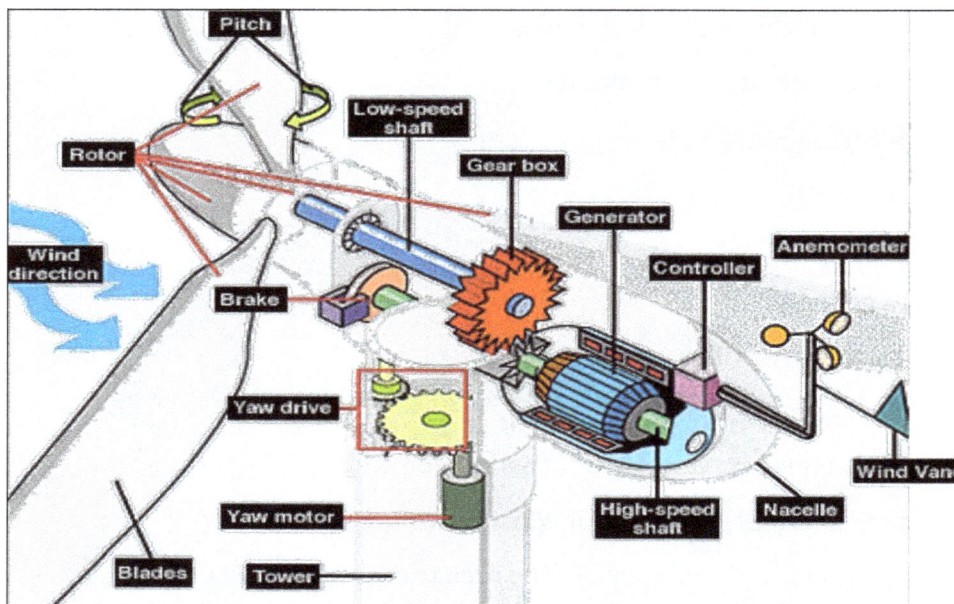

Schematic of wind turbine components.

Larger sizes are mandated by two reasons. They are cheaper and they deliver more energy. Their energy yield is improved partly because the rotor is located higher from the ground and so intercepts higher velocity winds, and partly because they are more efficient. The productivity of the 600 kW machines is around 50 percent higher than that of the 55 kW machines. Reliability has improved steadily with wind turbine manufacturers guaranteeing availabilities of 95 percent.

Wind energy systems include the following major components:

The rotor and its blades, the hub assembly, the main shaft, the gear box system, main frame, transmission, yaw mechanism, overspeed protection, electric generator, nacelle, yaw drive, power conditioning equipment, and tower.

### Wind Turbine Aerodynamics

The wind turbine rotor interacts with the wind stream, resulting in a behaviour named aerodynamics, which greatly depends on the blade profile.

## Actuator Disc Concept

The analysis of the aerodynamic behaviour of a wind turbine can be done, in a generic manner, by considering the extraction process.

Consider an actuator disc and an air mass passing across, creating a stream-tube.

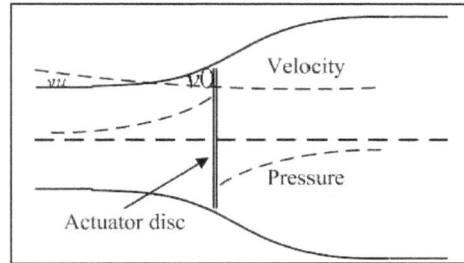

Energy extracting actuator disc.

The conditions (velocity and pressure) in front of the actuator disc are denoted with subscript u, the ones at the disc are denoted with o and, finally, the conditions behind the disc are denoted with w. The momentum H m ($v_u - v_w$) transmitted to the disc by the air mass m passing through the disc with crosssection A produces a force, expressed as:

$$T = \frac{\Delta H}{\Delta t} = \frac{\Delta m (v_u - v_w)}{\Delta t} = \frac{\rho A v_0 \Delta t (v_u - v_w)}{\Delta t} = \rho A v_0 (v_u - v_w)$$

$$T = A\left(p_0^+ - p_0^-\right)$$

Using Bernoulli's equation, the pressure difference is:

$$p_0^+ - p_0^- = \frac{1}{2}\rho\left(v_u^2 - v_w^2\right)$$

and, replacing Equations results in:

$$T = \frac{1}{2}\rho A\left(v_u^2 - v_w^2\right)$$

From Equations the kinetic energy of an air mass travelling with a speed v is:

$$E_k = \frac{1}{2}mv^2,$$

where m is the air mass that passes the disc in a unit length of time, e.g., $m = \rho A v_0$ then the power extracted by the disc is:

$$P = \frac{1}{2}\rho A v_0\left(v_u^2 - v_w^2\right)$$

The power coefficient, denoting the power extraction efficiency, is defined as:

$$P = \frac{1}{2}\rho A v^3 \, 4a(1-a)^2,$$

$$C_p = \frac{P}{P_t} = \frac{0.5.\rho A v^3 . 4a(1-a)^2}{0.5.\rho A v^3}$$

$$C_p = 4a(1-a)^2$$

The maximum value of $C_p$ occurs for a 1 3 and is $C_{pmax}$ 0.59, known as the Betz limit and represents the maximum power extraction efficiency of a wind turbine.

## Power Obtained from Wind

A wind turbine obtains its power input by converting the force of the wind into torque (turning force) acting on the rotor blades. The amount of energy which the wind transfers to the rotor depends on the density of the air, the rotor area, and the wind speed.

Density of air: The kinetic energy of a moving body is proportional to its mass. The kinetic energy in the wind thus depends on the density of the air, i.e. its mass per unit of volume.

In other words, the "heavier" the air, the more energy is received by the turbine. At normal atmospheric pressure and at 15°C, the density of air is 1.225 kg/m³, which increases to 1.293 kg/m³ at 0°C and decreases to 1.164 kg/m³ at 30°C. In addition to its dependence upon temperature, the density decreases slightly with increasing humidity. At high altitudes (in mountains), the air pressure is lower, and the air is less dense. It will be shown later in this chapter that energy proportionally changes with a variation in density of air.

Rotor area: When a farmer tells how much land he is farming, he will usually state an area in terms of square meters or hectares or acres. With a wind turbine it is much the same story, though wind farming is done in a vertical area instead of a horizontal one. The area of the disc covered by the rotor (and wind speeds, of course), determines how much energy can be harvested over a year.

A typical 1,000 kW wind turbine has a rotor diameter of 54 m, i.e. a rotor area of some 2,300 m². The rotor area determines how much energy a wind turbine is able to harvest from the wind. Since the rotor area increases with the square of the rotor diameter, a turbine which is twice as large will receive 2², i.e. four times as much energy.

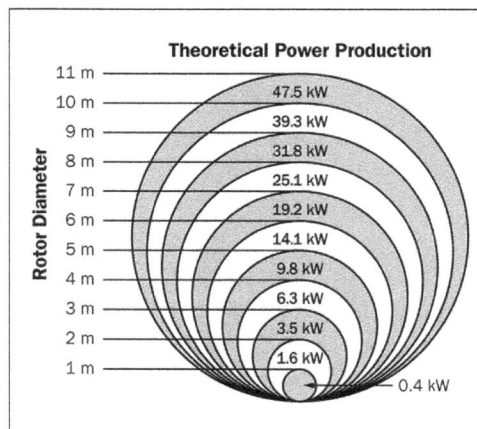

**Theoretical Power Production**

Rotor diameters may vary somewhat from the figures given above, because many manufacturers optimize their machines to local wind conditions: A larger generator, of course, requires more power (i.e. strong winds) to turn at all. So if one installs a wind turbine in a low wind area, annual output will actually be maximized by using a fairly small generator for a given rotor size (or a larger rotor size for a given generator). The reason why more output is available from a relatively smaller generator in a low wind area is that the turbine will be running more hours during the year.

Wind velocity: Considering an area A (e.g. swept area of blades) and applying a wind velocity v, the change in volume with respect to the length "l" is:

$$V = A \cdot l, \; V = l / t$$
$$V = A \cdot v \cdot t,$$

The energy in the wind is in the form of kinetic energy. Kinetic energy is characterized by the equation:

$$E = 1 / 2mv^2$$

The change in energy is proportional to the change in mass, where:

$$m = V \cdot \rho a$$

and ρa the specific density of the air. Therefore, substituting for V and m yields:

$$E = 1 / 2^* A^* \rho a^* v^{3\,*} t$$

## Energy Conversion in Wind

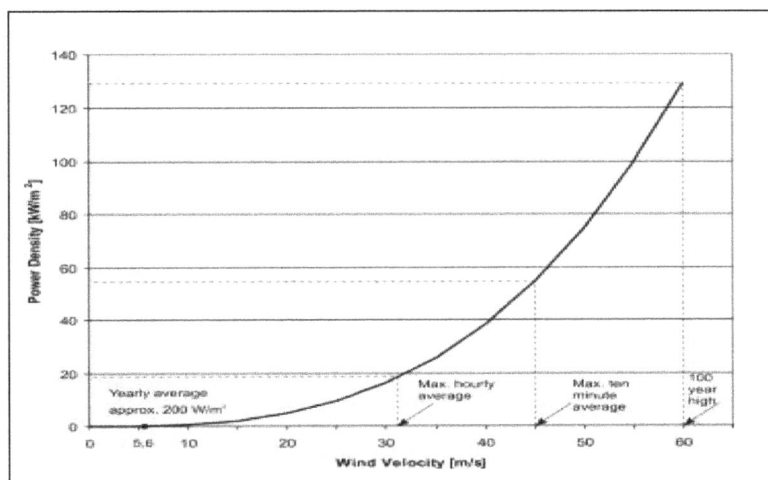

Relationship between wind velocity and power of wind (wind speed for Germany).

From the previous equation it can be seen that the energy in the wind is proportional to the cube of the wind speed, v $^3$.

Therefore, power in wind is proportional to v$^3$. It can be seen that the power output per m² of the rotor blade is not linearly proportional to the wind velocity, as proven in the theory above. This means that it is more profitable to place a wind energy converter in a location with occasional high winds than in a location where there is a constant low wind speed.

Measurements at different places show that the distribution of wind velocity over the year can be approximated by a Weibull-equation. This means that at least about 2/3 of the produced electricity will be earned by the upper third of wind velocity. From a mechanical point of view, the power density range increases by one thousand for a variation of wind speed of factor 10, thus producing a construction limit problem. Therefore, wind energy converters are constructed to harness the power from wind speeds in the upper regions.

## Momentum Theory

Compared to the Rankine–Froude model, Blade element momentum theory accounts for the angular momentum of the rotor. Consider the left hand side of the figure below. We have a stream tube, in which there is the fluid and the rotor. We will assume that there is no interaction between the contents of the stream tube and everything outside of it.

That is, we are dealing with an isolated system. In physics, isolated systems must obey conservation laws. An example of such is the conservation of angular momentum. Thus, the angular momentum within the stream tube must be conserved. Consequently, if the rotor acquires angular momentum through its interaction with the fluid, something else must acquire equal and opposite angular momentum. As already mentioned, the system consists of just the fluid and the rotor, the fluid must acquire angular momentum in the wake.

As we related the change in axial momentum with some induction factor, we will relate the change in angular momentum of the fluid with the tangential induction factor Let us consider the following setup. We will break the rotor area up into annular rings of infinitesimally small thickness. We are doing this so that we can assume that axial induction factors and tangential induction factors are constant throughout the annular ring. An assumption of this approach is that annular rings are independent of one another i.e. there is no interaction between the fluids of neighboring annular rings.

## Bernoulli for Rotating Wake

Let us now go back to bernoulli:

$$\frac{1}{2}\rho v_1^2 + P_1 = \frac{1}{2}\rho v_2^2 + P_2$$

The velocity is the velocity of the fluid along a streamline. The streamline may not necessarily run parallel to a particular co-ordinate axis, such as the z-axis. Thus the velocity may consist of components in the axes that make up the co-ordinate system. For this analysis, we will use cylindrical polar co-ordinates $(r,\theta,z)$. Thus, $v^2 = v_r^2 + v_\theta^2 + v_z^2$.

## Pre-rotor

$$P_\infty + \frac{1}{2}\rho v_u^2 = P_D + \frac{1}{2}\rho v_D^2$$

where Vu is the velocity of the fluid along a streamline far upstream, and is VDthe velocity of the fluid just prior to the rotor. Written in cylindrical polar co-ordinates, we have the following expression:

$$P_\infty + \frac{1}{2}\rho v_\infty^2 = P_{D+} + \frac{1}{2}\rho\left(v_\infty\left(1-a\right)\right)^2$$

where V∞ and V∞ (1-α) are the z-components of the velocity far upstream and just prior to the rotor respectively. This is exactly the same as the upstream equation from the Betz model.

It should be noted that, as can be seen from the figure above, the flow expands as it approaches the rotor, a consequence of the increase in static pressure and the conservation of mass. This would imply that Vr≠0 upstream. However, for the purpose of this analysis, that effect will be neglected.

## Post-rotor

$$P_{D-} + \frac{1}{2}\rho v_D^2 = P_\infty + \frac{1}{2}\rho v_w^2$$

where VD, z=v∞(1-α)is the velocity of the fluid just after interacting with the rotor. This can be written as VD2= VD,r² + VD,θ²+ VD,z² The radial component of the velocity will be zero; this must be true if we are to

$$P_{D-} + \frac{1}{2}\rho v_{D,z}^2 = P_\infty + \frac{1}{2}\rho v_{w,z}^2 = P_{D-} + \frac{1}{2}\rho\left(v_\infty\left(1-a\right)\right)^2$$

In other words, the Bernoulli equations up and downstream of the rotor are the same as the Bernoulli expressions in the Betz model. Therefore, we can use results such as power extraction and wake speed that were derived in the Betz model i.e.

$$v_{w,z} = \left(1-2a\right)v_\infty$$

$$\text{Power} = 2a\left(1-a\right)^2 v_\infty^3\,\rho A_D$$

This allows us to calculate maximum power extraction for a system that includes a rotating wake. This can be shown to give the same value as that of the Betz model i.e. 0.59.

## Sabinin's Theory

The impulse theory as applied to the horizontal shaft wind turbine.

The application of the impulse theory to the horizontal shaft wind turbine dimensioning has been taken over from the airplane propeller elastic theory, which, together with the classical whirlwind theory has led to highly efficient aerodynamic solutions.

One of the forerunners of this calculation and dimensioning method was G. K. Sabinin who had his studies in this field published starting from 1923.

## Hypotheses

The application of the impulse theory to the calculation and dimensioning of the propeller – type horizontal shaft turbines is based on the following main hypotheses:

- The air jet crosses the rotor at an even velocity throughout the axial cross section.

- The rotor lets the air pass through the blades without determining a local velocity discontinuity and has an infinite number of blades.

- The presence of the rotor brings about a pressure variation between upstream and downstream, in a fluid domain delimited downstream the catching system by a cylindrical surface on which an infinite number of whose winding is the cylindrical surface corresponding to section A-A, downstream, in its immediate vicinity.

- The current tube delimited by the solenoid surface does not allow for the air exchange between its inside and outside, the air current passing through the rotor being considered isolated from the ambient.

- The current non-uniformity increases at the rotor outlet, which leads to turbulent energy losses and therefore to a lower efficiency of catching the wind energy, while the current twisting dissipated in alternating whirlpools caused by the instability of the flow downstream the rotor.

- Air pressure in the 0 – 0 cross-section is assumed to be equal to the atmospheric one.

- In cross-section 1-1, pressure rises to values $p_1 < p_0$, yet further increasing while aiming asymptotically towards values $p_0$, far-off downstream.

- In the sections downstream the wind turbine rotor, pressure variation is neglected, considering that the centrifugal forces determined by the refined current twisting after rotor crossing are small as compared to the forces caused by the axial impulse in the same section.

- The pressure difference downstream and upstream the catching system leads to the occurrence of an axial force upon the rotor "Fa"; as a result of the adequately built rotor geometry there appears a tangential component "Fr" in the rotation plan, leading to the occurrence of the catching system useful moment. Obviously, wind-catching systems should be builder so that the rotor impulse tangential component should be as big as possible.

The air current shape upstream and downstream the rotor according to the above hypotheses, as well as the diagram of the modality the whirlwind solenoid surface is formed, delimiting the current tube downstream the rotor; upstream the rotor, current velocity $V_0$ is equal to the far-off upstream velocity (infinite upstream); in the catching system rotor section 1-1, and as getting nearer and nearer to this topic, the current axial velocity drops to values $V_{11} = V_0 - V_2$, where $V_1$ and $V_2$ are velocities induced by the cylindrical turbulent layer generated by the whirlwinds on top of the rotor blades forming up a current tube.

As in the wind turbine rotor there appears a rotating moment in the section 1-1, that leads to the occurrence of the induced rotating impulse, counterclockwise to the catching system rotation. there results therefore that downstream the rotor, the air current rotates at a velocity equal to the velocity V2 in a close enough section, where the whirlwind tubes do not vicinity, the current peripheral velocity is considered not very much different from that, so that V2 ~ V2_, where V2 is the rotating velocity at the rotor outlet.

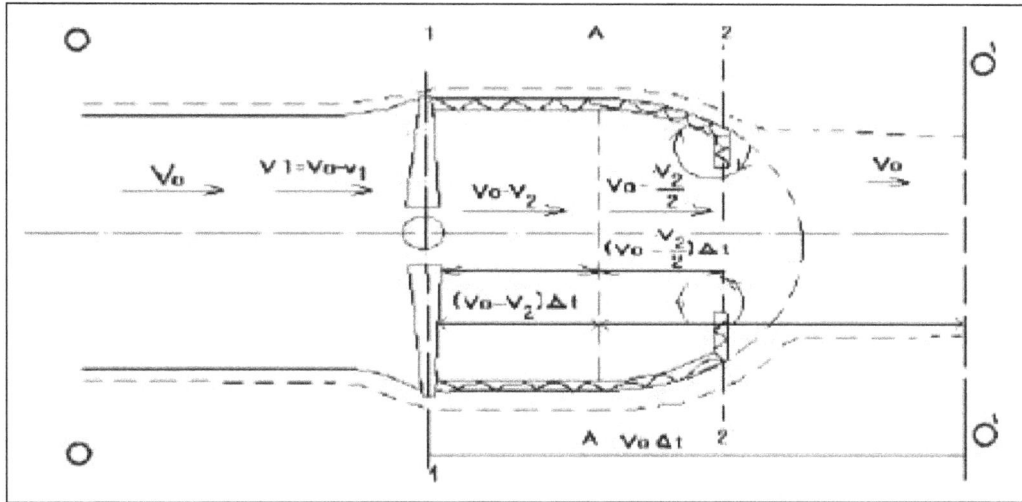

Formation of whirlwind solenoid surface Downstream the wind turbine.

Determining of the peripheral force component and of the axial force component acting upon the catching system blade Isolating an annular area of r radius, dr thick, off the air current, the axial component dFa as well as the rotation peripheral component dFr can be determined in the corresponding annular section. These components lead to the occurrence of the interactions in the rotor construction elements (blades). By applying the impulse theorem and taking into account that, in keeping with second principle of mechanics, action is equal to reaction, there can be determined the reactions occurring in the rotor blades, along the axial and tangential directions, respectively,

$$dm = 2\pi \cdot r \cdot p \cdot V1 \cdot dr$$

Where, dFa – elementary axial force, and dFr – the elementary tangential force.

Where, Dm – is the unitary mass ir flow crossing the annular section in the rotor.

$V_1 = V_0 - V_1$; $V_1$ is the induced velocity before the rotor, caused by the blades. By applying to the air mass delimited by the two concentric cylindrical areas of radius r and r + dr,

$$dFa = V_2 \, dm;$$

In a similar way, in keeping with theorem of the motional quantity moment, the elementary torque created by the tangential rotational force occurring on the elementary surface of the blades enclosed between the two cylindrical areas of radius r and r+dr, there results the relation:

$$dFr = U_2 dm$$

V1, u1, w1, as well the corresponding velocities in a section downstream the rotor are written with the following relations: $V_2 = -V_1$, of the blades enclosed between the two cylindrical areas of radius r and r+dr, there results the relation:

$$u_2 = {}_\tau u_1$$

$$\frac{v^2}{u^2} = \frac{(v_0 - v_1)\mu + (\omega r + u_1)}{(v_0 - v_1) - \mu(\omega r + u_1)}$$

The relation represents the ratio of the axial induced velocity to the peripheral velocity component us. This ratio is variable along the blade from hub to apex. Relation is fulfilled provided what is in between the brackets is equal to zero. Out of this condition, there result:

$$v^2 \left( v_1 - \frac{v_2}{2} \right) = u_2 \left( \frac{u_2}{2} - u_1 \right) \qquad v_2 = \frac{v_2}{2} \text{ And } \quad u_2 = \frac{u_2}{2}$$

## Variable Speed Systems

### Need of Variable Speed Systems

There are many similarities in major components construction of fixed-speed wind turbines and wind turbines operating within a narrow variable-speed range. Fixed-speed wind turbines operating within a narrow speed range usually use a double-fed induction generator and have a converter connected to the rotor circuit. The rotational speed of the double-fed induction generator equally 1000 or 1500 rpm, so a gearbox implementation the required.

To simplify the nacelle design a direct-driven generator is used. A direct-driven generator using a large turbine blades diameter can operate at a very low speeds and does not need a gearbox installed to increase to speed.

The usage of frequency converter is needed to use a direct-driven generator, so wind turbines operating within a broad variable-speed range are equipped with a frequency converter. In an conventional fixed-speed wind turbine, the gearbox and the generator have to be mounted on a stiff bed plate and aligned precisely in respect to each other. A direct driven generator can be integrated with the nacelle, so the generator housing and support structure are also the main parts of the nacelle construction.

Nowadays, many wind turbines manufacturers are using variable speed wind turbine systems. The electrical system for variable speed operation is a lot more complicated, in comparison to fixed speed wind turbine system. The variable speed operation of a wind turbine can be obtained in many different ways with differentiation for a broad or a narrow wind speed range.

The main difference between wide and narrow wind turbines speed range is the energy production and the capability of noise reduction. A broad speed range gives larger power production and causes reduction of the noise in comparison to a narrow speed range system. One of the biggest advantages of variable speed systems controlled in a proper way is the reduction of power fluctuations emanating from the tower shadow.

## Power Wind Speed Characteristics

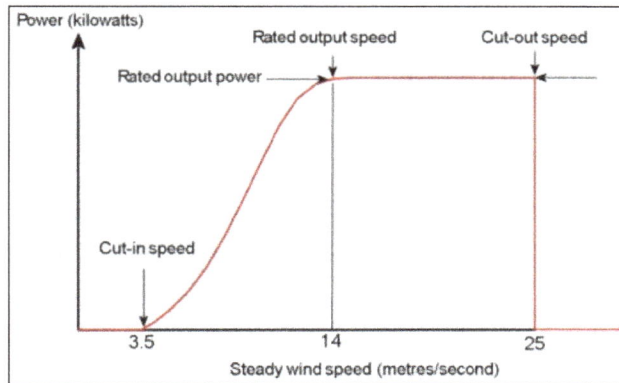

Typical wind turbine power output with steady wind speed.

## Cut-in Speed

At very low wind speeds, there is insufficient torque exerted by the wind on the turbine blades to make them rotate. However, as the speed increases, the wind turbine will begin to rotate and generate electrical power. The speed at which the turbine first starts to rotate and generate power is called the cut-in speed and is typically between 3 and 4 metres per second.

## Rated Output Power and Rate Output Wind Speed

As the wind speed rises above the cut-in speed, the level of electrical ouput power rises rapidly as shown. However, typically somewhere between 12 and 17 metres per second, the power output reaches the limit that the electrical generator is capable of. This limit to the generator output is called the rated power output and the wind speed at which it is reached is called the rated output wind speed. At higher wind speeds, the design of the turbine is arranged to limit the power to this maximum level and there is no further rise in the output power. How this is done varies from design to design but typically with large turbines, it is done by adjusting the blade angles so as to to keep the power at the constant level.

## Cut-out Speed

As the speed increases above the rate output wind speed, the forces on the turbine structure continue to rise and, at some point, there is a risk of damage to the rotor. As a result, a braking system is employed to bring the rotor to a standstill. This is called the cut-out speed and is usually around 25 metres per second.

## Wind Turbine Efficiency or Power Coefficient

The available power in a stream of wind of the same cross-sectional area as the wind turbine can

$$C_P = \frac{P}{P_{in}} = \frac{t^{\frac{1}{3}}(1-a2}{\frac{1}{2}\rho Av \, \frac{3}{t \, 1}} = \frac{1}{2}(1-a^2)(1+a)$$

If the wind speed U is in metres per second, the density p is in kilograms per cubic metre and the in rotor diameter d is in metres then the available power is watts. The efficiency, μ, or, as it is more commonly called, the power coefficient, cp, of the turbine is simply defined as the actual wind power delivered divided by the available power.

## The Betz Limit on Wind Turbine Efficiency

There is a theoretical limit on the amount of power that can be extracted by a wind turbine from an airstream. It is called the Betz limit and the proof of this limit is given. The limit is

$$\mu = 16/27 \approx 59\%$$

## Variable Speed Constant Frequency Systems

The evolution of wind power conversion technology has led to the development of different types of wind turbine configurations that make use of a variety of electric generators. A classification of most common electric generators in large wind energy conversion systems (WECS) is presented in figure.

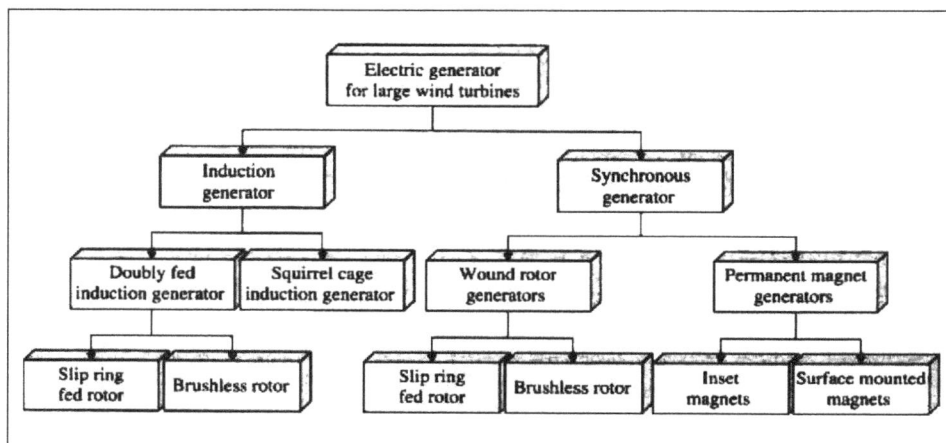

Classification of commonly used electric generators in large wind turbines The wound-rotor induction generator, also known as the doubly fed induction generator (DFIG), is one of the most commonly used generators in the wind energy industry. The wound-rotor synchronous generator (WRSG) is also found in practical WECSs with high numbers of poles operating at low rotor speeds. Squirrel-cage induction generators (SCIGs) are also widely employed in wind energy systems where the rotor circuits (rotor bars) are shorted internally and therefore not brought out for connection with external circuits. In permanent-magnet synchronous generators (PMSGs), the rotor magnetic flux is generated by permanent magnets. Two types of PMSG are used in the wind energy industry: surface mounted and inset magnets.

## Synchronous Generators

In wind energy application synchronous generators are used in variable speed systems only. The converter to decouple the frequency between machine and grid has to be designed for full load, different from the doubly-fed induction generator concept. Generators of larger ratings are generally equipped with an excitation winding, fed via slip-rings from a separate exciter.

Synchronous machines are more suitable for designs with large pole numbers than induction machines. Hence they are the option for direct driven generators, sparing the gear box in the system. Generators of considerable diameter and pole number values are found in the gearless Magawatt systems.

The conventional synchronous generator can be used with a very cheap and efficient diode rectifier. The synchronous generator is more complicated than the induction generator and should therefore be somewhat more expensive. However, standard synchronous generators are generally cheaper than standard induction generators. A fair comparison cannot be made since the standard induction generator is enclosed while the synchronous generator is open-circuit ventilated. The low cost of the rectifier as well as the low rectifier losses make the synchronous generator system probably the most economic one today. The drawback of this generator and rectifier combination is that motor start of the turbine is not possible by means of the main frequency converter.

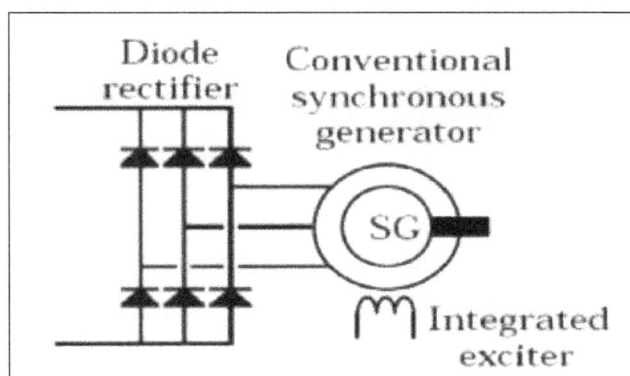

Variable Speed Systems Synchronous generator.

## Synchronous Generator and Diode-thyristor Converter

The generator system is consisting of a synchronous generator, a diode rectifier, a dc filter and a thyristor inverter. The inverter may have a harmonic filter on the network side if it is necessary to comply with utility demands. The harmonic filter is, however, not included in the efficiency calculations in this report.

The advantage of a synchronous generator is that it can be connected to a diode or thyristor rectifier. The low losses and the low price of the rectifier make the total cost much lower than that of the induction generator with a self-commutated rectifier. When using a diode rectifier the fundamental of the armature current has almost unity power factor. The induction generator needs higher current rating because of the magnetization current. The disadvantage is that it is not possible to use the main frequency converter for motor start of the turbine. If the turbine cannot start by itself it is necessary to use auxiliary start equipment. If a very fast torque control is important, then a generator with a self-commutated rectifier allows faster torque response. A normal synchronous generator with a diode rectifier will possibly be able to control the shaft torque up to about 10 Hz, which should be fast enough for most wind turbine generator systems.

## Doubly Fed Induction Generators (DFIG)

The most common variable-speed wind turbines are the Doubly Fed Induction Generator (DFIG), which offers high efficiency over a wide range of wind speeds as well as the ability to supply power at a constant voltage and frequency while the rotor speed varies. This technology consists of a wound rotor induction generator and a back-to-back power converter placed into the rotor of the machine while the stator is directly connected to the grid.

The power converter allows for the machine to be controlled between sub-synchronous speed and super-synchronous speed (a speed higher than the synchronous speed), usually, a variation from 40 % to 30 % of synchronous speed is chosen. This converter capacity is designed to handle 20–30 % of the machine rate, which is beneficial both economically and technically. The rotor of the DFIG has a three-phase winding similar to the stator winding. The rotor winding is embedded in the rotor laminations but in the exterior perimeter. This winding is usually fed through slip-rings mounted on the rotor shaft. In DFIG wind energy systems, the rotor winding is normally connected to a power converter system that makes the rotor speed adjustable.

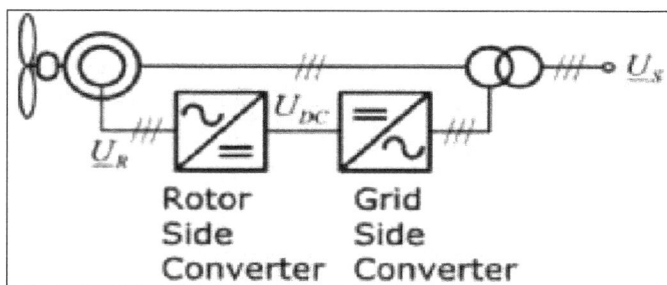

Double fed induction generator.

The fact that the rotor circuit of an SCIG is not accessible can be changed if the rotor circuit is wound and made accessible via slip rings, which offers the possibility of controlling the rotor circuit so that the operational speed range of the generator can be increased in a controlled manner. The rotor circuit is often connected to back-to-back power electronic converters, which consists of a rotor-side converter and a grid-side converter sharing the same DC bus, so that the difference between the mechanical speed of the rotor and the electrical speed of the grid can be compensated via injecting a current with a variable frequency into the rotor circuit. Hence, the operation during both normal and faulty conditions can be regulated by controlling the converters.

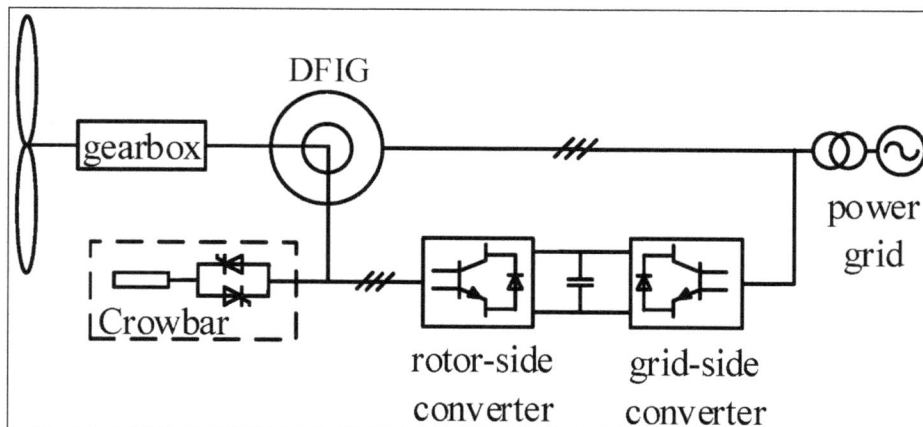

Doubly-fed induction generator wind turbine schematic.

The doubly fed induction machine (DFIM), doubly fed induction generator (DFIG), or wound rotor induction generator (WRIG) are common terms used to describe an electrical machine with the following characteristics:

- A cylindrical stator that has in the internal face a set of slots (typically 36–48), in which are located the three phase windings, creating a magnetic field in the air gap with two or three pairs of poles.

- A cylindrical rotor that has in the external face a set of slots, in which the three phase windings, are to chafed creating a magnetic field in the air gap of the same pair of poles as the stator.

- The magnetic field created by both the stator and rotor windings must turn at the same speed but phase shift to some degrees as a function of the torque created by the machine.

- As the rotor is a rotating part of the machine, to feed it, it's necessary to have three slip rings. The slip ring assembly requires maintenance and compromises the system reliability, cost, and efficiency.

A DFIG can be excited via the rotor windings and does not have to be excited via the stator windings. If needed, the reactive power needed for the excitation from the stator windings can be generated by the grid-side converter.

As a result, a wind power plant equipped with DFIGs can easily take part in the regulation of grid voltage. The stator always feeds real power to the grid but the real power in the rotor circuit can flow bidirectional, from the grid to the rotor or from the rotor to the grid, depending on the operational condition. Ignoring the losses, the power handled by the rotor circuit is:

$$P_{rotor} = -s \cdot P_{stator.}$$

Where s is the slip, and the power sent to the grid is:

$$P_{grid} = P_{rotor} + P_{stator} = (1-s) P_{stator}$$

Since most of the power flows through the stator circuit, the power processed by the rotor circuit can be reduced roughly to 30%. This means the great advantage of a sufficient range of operational speed can be achieved at a reasonably low cost.

DFIGs are often applied in variable speed wind turbine systems with a multi-stage gearbox. Its basic operating principle is the same as an SCIG-based system but the rotor active power is controlled by the power electronic converters so that a speed range of ±30% around the synchronous speed can be obtained. The choice of the rated power for the rotor converter is a tradeoff between cost and the desired speed range. Moreover, the converter compensates the reactive power and smooth's the grid connection.

Although a DFIG offers a sufficient range of operational speed and many other merits, it is very sensitive to voltage disturbances, especially voltage sags. Abrupt voltage drops at the terminals often cause large voltage disturbances on the rotor, which may exceed the voltage rating of the rotor- side converter (RSC), make the rotor current uncontrollable, and even damage the RSC. Many strategies are available to improve the low-voltage ride-through capability of DFIGs.

## Squirrel-cage Induction Generators (SCIG)

Squirrel-cage induction generators are exclusively used in the fixed-speed WECS. The generator power rating is in the range of a few kilowatts to a few megawatts. For large wind farms, megawatt generators are widely employed. The squirrel-cage induction generator is simple, reliable, cost-effective, and maintenance-free compared to other types of wind generators. This is achieved by using a robust squirrel-cage rotor structure, in which the rotor winding is made of copper bars embedded in the rotor magnetic core. As a result, slip rings and brushes that are required in the wound-rotor induction and synchronous generators are eliminated.

There are two main types of induction generators in the wind energy industry: doubly fed induction generators (DFIGs) and squirrel-cage induction generators (SCIGs). These generators have the same stator structure and differ only in the rotor structure. Figure shows the construction of a squirrel-cage induction generator. The stator is made of thin silicon steel laminations. The laminations are insulated to minimize iron losses caused by induced eddy currents. The laminations are basically flat rings with openings disposed along the inner perimeter of the ring. When the laminations are stacked together with the openings aligned, a canal is formed, in which a three-phase copper winding is placed.

There are two main types of induction generators in the wind energy industry: doubly fed induction generators (DFIGs) and squirrel-cage induction generators (SCIGs). These generators have the same stator structure and differ only in the rotor structure. Figure shows the construction of a squirrel-cage induction generator. The stator is made of thin silicon steel laminations. The laminations are insulated to minimize iron losses caused by induced eddy currents. The laminations are basically flat rings with openings disposed along the inner perimeter of the ring. When the laminations are stacked together with the openings aligned, a canal is formed, in which a three-phase copper winding is placed.

The rotor of the SCIG is composed of the laminated core and rotor bars. The rotor bars are embedded in slots inside the rotor laminations and are shorted on both ends by end rings. When the stator winding is connected to a three-phase supply, a rotating magnetic field is generated in the air gap. The rotating field induces a three-phase voltage in the rotor bars. Since the rotor bars are shorted, the induced rotor voltage produces a rotor current, which interacts with the rotating field to produce the electromagnetic torque.

A simplified diagram of the induction generator is shown in figure, where the multiple coils in the stator and multiple bars in the rotor are grouped and represented.

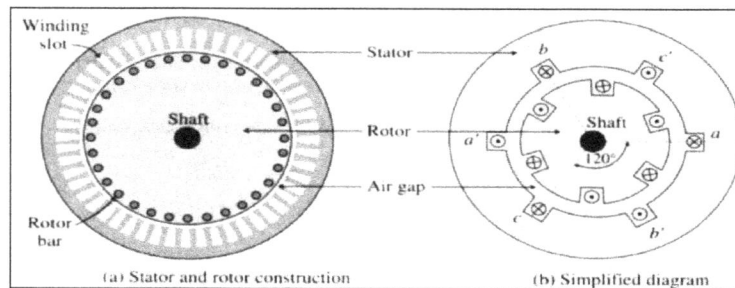

Simplified diagram of the induction generator.

There are two commonly used dynamic models for the induction generator. One is based on space vector theory and the other is the dq-axis model derived from the space vector model. The space vector model features compact mathematical expressions and a single equivalent circuit but requires complex (real and imaginary part) variables, whereas the dq-frame model is composed of two equivalent circuits, one for each axis.

These models are closely related to each other and are equally valid for the analysis of transient and steady-state performance of the induction generator. A squirrel-cage induction machine is often operated as a motor but it can be operated as a generator when driven by a prime mover to a speed exceeding the synchronous speed. Induction machines are widely applied as generators in wind power applications due to the reduced unit cost and size, ruggedness, lack of brushes, absence of a separate DC source, ease of maintenance, self-protection against severe overloads and short circuits, etc.

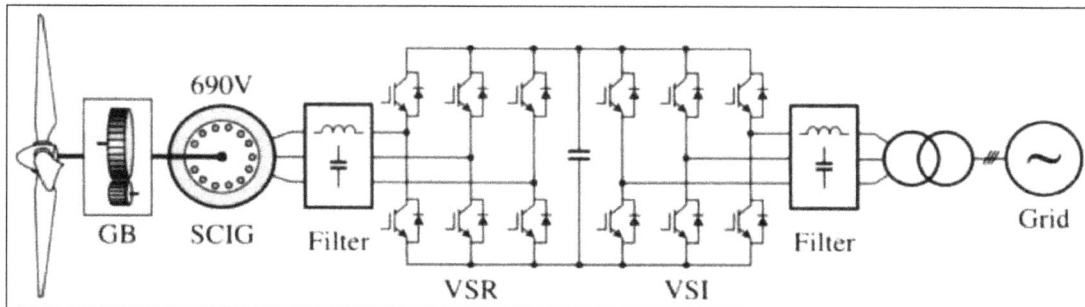

Squirrel-cage induction generator WECS.

An induction generator produces real power but it needs reactive power to establish the excitation (the magnetic field). This leads to a low power factor, which is often penalized by utility companies. The reactive power needed for excitation can be provided by a capacitor bank, the grid or a solid-state power electronic converter. The connection of an SCIG, in particular a big one, to the grid often causes a large inrush current that is 7 ~ 8 times of the rated current and a soft-starter is often needed. The pole pair number of SCIG used in commercial fixed-speed wind turbines is often equal to 2 or 3, which corresponds to a synchronous speed of 1500 rpm or 1000 rpm for a 50 Hz system. As a result, a three-stage gearbox is often required in the drive train. SCIGs are often applied in fixed-speed wind turbine systems directly connected to the grid through a transformer, as shown in figure.

The need for a three-stage gearbox in the drive train considerably increases the weight of the nacelle, and the investment and maintenance costs. Moreover, it is necessary to obtain the excitation current from the grid, which makes impossible to support the grid voltage.

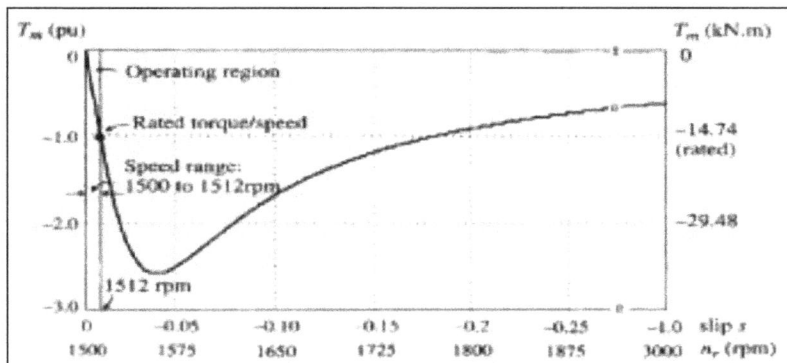

Torque Vs Speed curve of squirrel-cage induction generator WECS.

## Variable Speed Generators Modelling

The complete generator system and its main components are shown in figure. The turbine is described by its power Pt and speed nt. The speed is raised to the generator speed ng via a gear. Pg is the input power to the generator shaft. The generator can be magnetized either directly by the field current If fed from slip rings or by the exciter current IE. The exciter is an integrated brushless exciter with rotating rectifier. The output electrical power from the generator armature is denoted by Pa. The generator armature current Ia and voltage Ua are rectified by a three-phase diode rectifier.

The rectifier creates a dc voltage Udr and a dc current Idr. On the other side of the dc filter the inverter controls the inverter dc voltage Udi and dc current Idi. Ud is the mean dc voltage and Id is the mean dc current. The power of the dc link Pd is the mean value of the dc power, equal to Id Ud. The inverter ac current is denoted Ii and the inverter ac voltage Ui. The ac power from the inverter is denoted Pi.

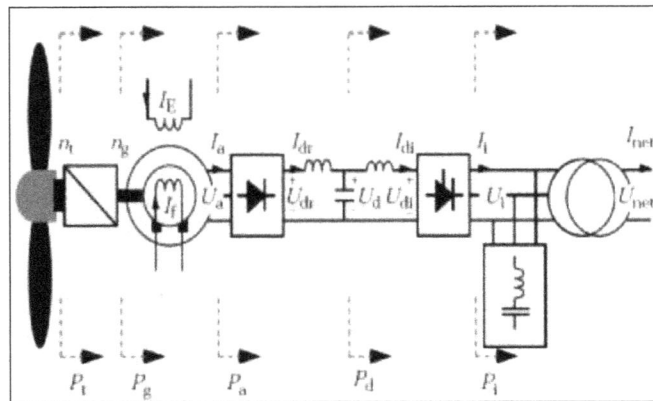

The total system and the quantities used. The generator can be magnetized either by slip rings or by an integrated exciter.

The filter is used to take care of the current harmonics by short circuiting the major part. The output of the generator system is the network current Inet. The network voltage is denoted Unet.

## Drive Selection

The variable-speed operation can capture theoretically about one-third more energy per year than the fixed-speed system. The actual improvement reported by the variable-speed systems operators in the field is lower, around 20 to 35 percent. However, the improvement of even 15 to 20 percent in the annual energy yield by variable-speed operation can make the systems commercially viable in low wind region. This can open a whole new market for the wind-power installations, and this is happening at present in many countries. Therefore the newer installations are more likely to use the variable- speed systems. As of 1997, the distribution of the system design is 35 percent one fixed speed, 45 percent two fixed-speed and 20 percent variable-speed power electronics systems. How waver the market share of the variable-speed systems, however, is increasing every year.

## Variable Speed Variable Frequency Schemes (VSVFS)

This figure is suitable for loads that are frequency insensitive such as heating load.

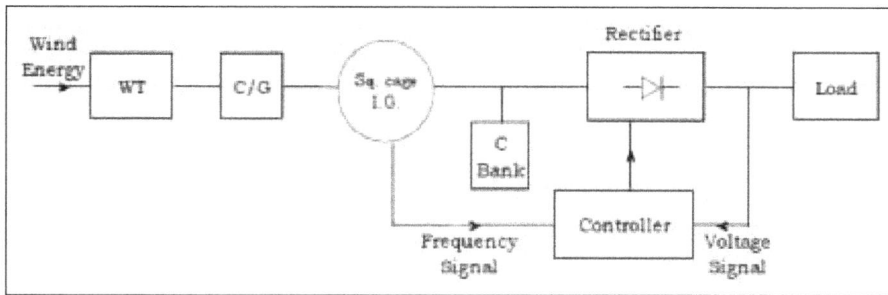

Depending upon the wind speed, squirrel cage Induction Generator generates power at variable frequency. Such generators are excited by Capacitor-bank. The magnitude and frequency of the generated emf depends upon the wind turbine speed, excitation capacitance and load impedance. If load requires constant dc voltage, output of generators is converted into d.c. using chopper controlled rectifiers. Feedback system can be used to monitor and control to get desired performance.

Advantages of the Variable Speed WECS with Respect to the Constant Speed WECS.

- For the same turbine WECS variable speed allows higher power capture, thereby increasing the annual energy output significantly.

- The variable speed WECS is capable of providing the required reactive power of the induction generator from the dc bus capacitance.

- Variable speed operation also allows a standard single winding machine to be used over the entire operating range of the turbine.

- In variable speed WECS since torque of the machine is controlled (either by field-orientation or direct torque control) the generator cannot be overloaded at any point of time beyond the prescribed limits.

Disadvantages of Variable Speed WECS with respect to the Constant Speed WECS.

- The power rating of the generator in the variable speed scheme should be five times greater than that of the optimal version of the constant speed case.

- Operating the generator over a wide speed range may result in a considerable reduction in the overall efficiency of the energy conversion process.

# Grid Connected Systems

Wind Interconnection Requirements:

- On and operation for generating plants.

- In general, applicable for all kind of generating plants as well as WPP's.

- Requirements – apply normally at a defined reference point of the generating plant.

Connection requirements can be split into different groups:

- General requirements (e.g. definitions, MW size limits, reference points).

- Required on steady state operation (e.g. operation ranges, power quality). Required on dynamic performance - different control schemes - during grid faults.

- Communication (e.g. SCADA), protection and verification (e.g. protection settings at WT, WPP level).

- Required on simulation models/validation.

- Compliance of WPP with requirements (type tests/certification, WPP compialce tests/ monitoring).

- "Additional requirement".

Interconnection Requirements – Challenges:

- Problems for the wind industry (e.g. manufacturers, developers, operators).

- Requirements changing quite frequently (e.g. updates and drafting of new rules). Requirements are diverse and sometimes that contain technical gray zones.

- It must be fully clear what is required and what has to be fulfilled by the WPP at which reference point.

- A common specification language is missing.

## Low-voltage Ride through (LVRT)

LVRT (Low Voltage Ride Through – also known as FRT - Fault Ride Through) has become a crucial feature of the wind turbine control system. The LVRT-term is capturing the ability of a wind turbine (or in reality a wind park) to stay connected to the grid throughout a short mains voltage drop (a brownout) or a mains failure (a blackout).

When the voltage of the grid is dropping it is essential that a wind park stay online in order to prevent major blackouts. It is not only essential that the park stays online - it is equally essential that the park is working actively to compensate for the faulty grid condition.

In China major blackouts (as a result of entire wind parks tripping and getting offline as a result of a brownout) have been seen.

This has increased the focus of the LVRT feature of the wind turbine control system. It is a fact that many wind turbine control systems installed just a few years ago does not have the LVRT-feature - and cannot upgrade to an economical way- for the grid operator - acceptable performance.

That is why DEIF Wind Power Technology in close cooperation with the power converter supplier is enjoying a drastically increased interest for our control systems - not only for new installed turbines, but also for upgrading existing installed capacity. Laboratory testing as early as 2010 brought the proof-of-concept which has later been full-scale implemented.

We are proud of having got several wind parks controlled by our control solutions full-scale field tested and approved as being in full LVRT-feature compliance by Chinese grid-operators. Whether the LVRT capability of a wind farm is satisfying for meeting the requirements is defined in grid codes issued by the grid operator. The capability of meeting these demands is decisive for whether the wind turbine/the wind park is allowed to be connected to the grid.

Examples of LVRT demands to the wind park (and derived from that – to the individual wind turbines) are:

- For short system faults (lasting up to 140ms) the wind farm has to remain connected to the grid. For supergrid (hv-grids) voltage dips of longer durations the wind farm has to remain connected to the grid up to more than 3 minutes.

- During grid faults or brownouts a wind farm has to supply maximum reactive current to the grid without exceeding the transient rating of the plant.

- On super-grids during voltage dips lasting more than 140ms the active power output of a wind farm has to be retained at least in proportion to the retained balanced supergrid voltage.

As mentioned above the LVRT-demands are individually specified by the grid operators and might therefore vary from operator to operator and from country to country. For wind turbines the LVRT testing is described in the standard IEC 61400-21. the LVRT-feature of wind turbine controls from DEIF WPT in combination with our Park Power Management solutions (including our forecasting solution) is all a wind park owner need to be in perfect compliance with the demands of grid operators worldwide.

## Ramp Rate Limitations

The ramp rate of wind power output is defined as the power changes from minute to minute, so its unit is [MW/minute]. Some ramp events result in severe ramp rates of power generation that exceed a "ramp rate limit" (RRL), which represents the capability of the remaining power system to compensate for wind ramps.

Those ramp events, such as the sudden die-off and rise, not only disturb the balance of demand and supply, but also hamper the participation of wind power in the electricity market. Wind power curtailment and reserve services can reduce severe ramp events, but they waste potential wind power energy and increase the thermal generation costs, respectively. Recently, a battery has been used to limit severe ramp events.

A battery can be charged during ramp-up events and discharged when ramp-down events without either wasting energy (except for energy loss in round trip charging and discharging of the battery) or requiring reserve services.

The use of batteries can also help stakeholders maximize profits by arbitraging electricity prices. However, the power rating, battery capacity, and operation policies must be designed before establishing a battery system. Therefore, the goal of this paper is to design battery parameters in order to compensate almost all severe ramp events of wind power output.

## Supply of Ancillary Services for Frequency and Voltage Control

Controlling frequency and voltage has an essential part of operating a power system.

However, since the liberalization of the electricity supply industry, the resources required to achieve this control have been treated as services that the system operator has to obtain from other industry participants.

Because this liberalization has proceeded independently in different parts of the world and because of the structural differences in the underlying power systems, the technical definitions of these services and the rules governing their trading vary considerably.

## Frequency Control Services

Maintaining the frequency at its target value requires that the active power produced and consumed be controlled to keep the load and generation in balance. A certain amount of active power, usually called frequency control reserve, is kept available to perform this control. The positive frequency control reserve designates the active power reserve used to compensate for a drop in frequency. On the other hand, the deployment of negative frequency control reserve helps to decrease the frequency.

Three levels of controls are generally used to maintain this balance between load and generation. Primary frequency control is a local automatic control that adjusts the active power generation of the generating units and the consumption of controllable loads to restore quickly the balance between load and generation and counteract frequency variations.

In particular, it is designed to stabilize the frequency following large generation or load outages. It is thus indispensable for the stability of the power system. All the generators that are located in a synchronous zone and are fitted with a speed governor perform this control automatically. The demand side also participates in this control through the self-regulating effect of frequencysensitive loads such as induction motors or the action of frequency-sensitive relays that disconnect or connect some loads at given frequency thresholds.

However, this demand-side contribution is not always taken into account in the calculation of the primary frequency control response. The provision of this primary control is subject to some constraints. Some generating units that increase their output in response to a frequency drop which cannot sustain this response for an indefinite period of time.

Their contribution must therefore be replaced before it runs out. It is also important that the contributors to primary control be distributed across the interconnected network to reduce unplanned power transits following a large generation outage and enhance the security of the system. In addition, a uniform repartition helps to maintain the stability of islanded systems in case of a power system separation.

Secondary frequency control is a centralized automatic control that adjusts the active power production of the generating units to restore the frequency and the interchanges with other systems to their target values following an imbalance. In other words, while primary control limits stops frequency excursions, secondary control brings the frequency back to its target value. Only the generating units that are located in the area where the imbalance originated should participate in this control as it is the responsibility of each area to maintain its load and generation in balance.

Note that loads usually do not participate in secondary frequency controls. Contrary to primary frequency control, frequency secondary control is not indispensable. This control is thus not implemented in some power systems where the frequency is regulated using only automatic primary and manual tertiary control. However, secondary frequency control is used in all large interconnected systems because manual control does not remove overloads on the tie lines quickly enough. Within the UCTE, secondary frequency control is also called load-frequency control (LFC).

## Voltage Control Service

From a system perspective, the overall task of regulating the voltage is sometimes organized into a three-level hierarchy. Primary voltage control is a local automatic control that maintains the voltage at a given bus (at the stator in the case of a generating unit) at its set point. Automatic voltage regulators (AVRs) fulfill this task for generating units. Other controllable devices, such as static voltage compensators, can also participate in this primary control.

Secondary voltage control is a centralized automatic control that coordinates the actions of local regulators in order to manage the injection of reactive power within a regional voltage zone. This uncommon technology is used in France and Italy. Tertiary voltage control refers to the manual optimization of the reactive power flows across the power system.

In practice, because of the close link between voltage and reactive power in transmission networks, these three levels of control require that participating devices are able to generate or absorb reactive power. From the perspective of providers of voltage control services, it is convenient to divide the production of reactive power into a basic and an enhanced reactive power service. The basic or compulsory reactive power service encompasses the requirements that generating units must fulfill to be connected to the network.

The enhanced reactive power service is a non-compulsory service that is provided on top of the basic requirements. The terminology of voltage control is much more uniform than for frequency control and does not need to be discussed further.

## Current Practices and Industry Trends Wind Interconnection

Wind power has become the world's fastest growing renewable energy source. many benefits of the wind energy are environmental protection, economic development, diversity of the supply, rapid spread, transference and technological innovation, industrial scale electricity in network and the fact is that the wind does not pollute, it is abundant, free and unlimited. The world-wide wind power installed capacity has exceeded 120 GW and the new installation in 2008 alone was more than 27 GW. More than thousands of wind turbines operating, with a total nameplate capacity of 121,188 MW of which wind power in Europe accounts for 55%.

World wind generation capacity more than quadrupled between 2000 and 2006, doubling about every three years. 81% of wind power installations are in the US and Europe.

Wind power is often described as an "intermittent" energy source, and therefore unreliable. In fact, at power system level, wind energy does not start and stop at regular intervals, so the term "intermittent" is misleading. The output of aggregated wind capacity is variable, just as the power system itself is inherently variable. In the past, wind turbine generators were disconnected from the system during faults.

Nowadays, there is an increasing requirement for wind farms to remain connected to the power system during faults, since the wind power lost might affect the system stability Therefore, the wind turbine behavior during system performance and its influence in the system protection must be analyzed. One of the most frequent irrelevant features about integrating wind energy into the electricity network is that it is treated in isolation.

An electricity system in practice is modify like a massive bath tub, with hundreds of taps (power stations) providing the input and millions of plug holes (consumers) draining the output. The taps and plugs are always open and close. For the grid operators, the task is to make sure there is enough water in the tub to maintain system security. It is therefore the combined effects of all technologies, as well as the demand patterns, that matters. The specific nature of wind power as a distributed and variable generation source requires specific infrastructure investments and the implementation of new technology and grid management concepts.

High levels of wind energy in system can impact on grid stability, congestion management and transmission efficiency and transmission adequacy. A grid code covers all material technical aspects relating to connections to, and the operation and use of, a country's electricity transmission system. They lay down rules which define the ways in which generating stations connecting to the system must operate in order to maintain grid stability.

## Impact on Steady-state and Dynamic Performance of the Power System Including Modeling Issue

### Modeling for Steady State Analysis

The power system modeling including wind turbines for steady state analysis in PSS/E version 32 is fairly simple. Each individual wind turbine generator (WTG) is connected to a 690V bus and the WTGs are connected to the wind farm internal network through their 0.69/35 kV step-up transformers. The internal network is organized in eight rows or sections with five WTGs in each section. Within these rows, the wind turbines are connected through 35kV underground cables of different lengths and capacities depending on the location of each unit and the distance to the 35kV collector bus.

The load flow solution provides the initial conditions for subsequent dynamic simulations. The maximum and minimum limits of active and reactive power must be respected in order to achieve a successful initialization. Inconsistencies between the power flow and the dynamic model will result in an unacceptable initialization.

### Modeling for Dynamic Analysis

Initial dynamic model data file for the regional Power System is used for the study case. The dynamic data file so called dyr.file in PSS/E consist of dynamic parameter data for all conventional synchronous generators, turbines, exciters governors and other devices. The first step in dynamic simulation using initial dynamic file is to enter the detailed dynamic model data for Wind Farm, which is saved in a file.

This file contains a group of records, each of which defines the location of a dynamic WTG model in the grid along with the constant parameters of the model. The model includes generator, electrical control, wind turbine, and pitch control.

Dynamic simulation is performed based on the load flow data that provide the transmission grid, load, and generator data. In this study, a number of simulations are performed to investigate the WTGs model response subjected to grid disturbances. The most relevant disturbance for the study case is a three-phase symmetrical short-circuit fault on the 110kV interconnection bus. Additional fault events are simulated on the different busses of the Kosovo Transmission System to evaluate WTG dynamic pattern of behavior for various fault impedances.

## Power System Dynamic Performance and its Impacts

Power system's stability has been recognized as an important problem for secure system operation. Generally, transient stability is the main concern on the majority of the power systems. Aspower systems have evolved through continuing growth, new operation technologies and controls in highly stressed conditions have emerged. More precisely, voltage stability and frequency stability have become greater concerns than in the past. A clear understanding of different types of stability and how they are interrelated is essential for the satisfactory design and operation of power systems.

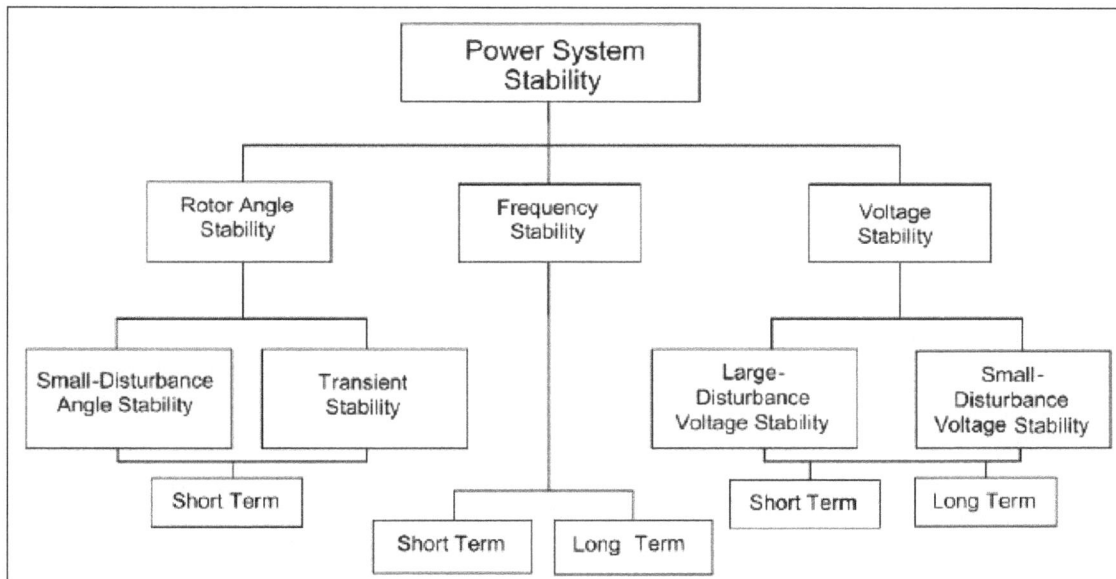

Categories of power system stabilities.

Power system stability is similar to the stability of any dynamic system, and has fundamental mathematical underpinnings. Precise definitions of stability can be found in the literature dealing with the rigorous mathematical theory of stability of dynamic systems. A circumstantial definition of Power System Dynamic Security and the corresponding types of Power System Stability is provided in.

Concise classification of power system stability is provided. Addionally, another significant issue is the relationship between the concepts of power system reliability, security, and stability of a power system.

Power System Reliability: It refers to the probability of a required operation condition achievement, with few interruptions over an extended time period.

Power System Security: It refers to the risk assessment of system ability to survive disturbances (taking into account the probability of these contingencies) without any power supply interruption.

Power System Stability: It refers to the dynamic operation after a severe or moderate disturbance, consequently to an initial steady state operating condition. The analysis of security relates to the determination of the power system robustness to imminent disturbances. Assuming that a power system subjected to changes, it is important that when the changes are completed, the system settles to a new operating state where no technical constraints are violated.

This implies that, in addition to the next operating conditions being acceptable, the system should survive the transition to these conditions. The above characterization of system security clearly highlights two aspects of its analysis:

- Static Security Analysis: It involves steady-state analysis of post-disturbance system conditions to verify that no voltage constraints are violated.

- Dynamic Security Analysis: It involves examining different categories of system stability described in section. The general practice for dynamic security assessment has been to use a deterministic approach.

The power system is designed and operated to withstand a set of contingencies selected on the basis that they have a significant possibility of occurrence. In practice, they are usually defined as the loss of any single element in a power system either spontaneously or preceded by a single, double, or three phase fault.

This method is generally called as the N-1 criterion because it examines the behavior of an N-component network following the loss of any one of its components. In addition, emergency controls, such as generation tripping, load shedding, and controlled islanding, may be used to withstand such events and prevent widespread blackouts.

Ongoing power systems under deregulated energy markets with a diversity of participants, the deterministic approach may not be appropriate. There is a need to account for the probabilistic nature of system conditions and events, and to quantify and manage risk. The trend will be to expand the use of risk-based security assessment. In this approach, the probability of the system becoming unstable and its consequences are examined, and the degree of exposure to system failure is estimated. This approach is computationally intensive but is possible with today's computing and analysis tools.

## Wind Energy Conversion System from Electrical Perspective

Green house gas reduction has been one of the crucial and inevitable global challenges, especially for the last two decades as more evidences on global warming have been reported. This has drawn increasing attention to renewable energies including wind energy, which is regarded as a relatively mature technology. It recorded 159 GW for the total wind energy capacities in 2009, which is the highest capacity among the existing renewable energy sources with excluding large-scale hydro power generators.

Also, its annual installation growth rate marked 31.7% in 2009 with its growth rate having been increasing for the last few years, which indicates that wind energy is one of the fastest growing and attractive renewable energy sources. The increasing price-competitiveness of wind energy against other conventional fossil fuel energy sources such as coal and natural gas is another positive indication on wind energy. Therefore, a vast amount of researches on WECS have been and is being undertaken intensively.

WECS consists of three major aspects; aerodynamic, mechanical and electrical.

The electrical aspect of WECS can further be divided into three main components, which are wind turbine generators (WTGs), power electronic converters (PECs) and the utility grid.

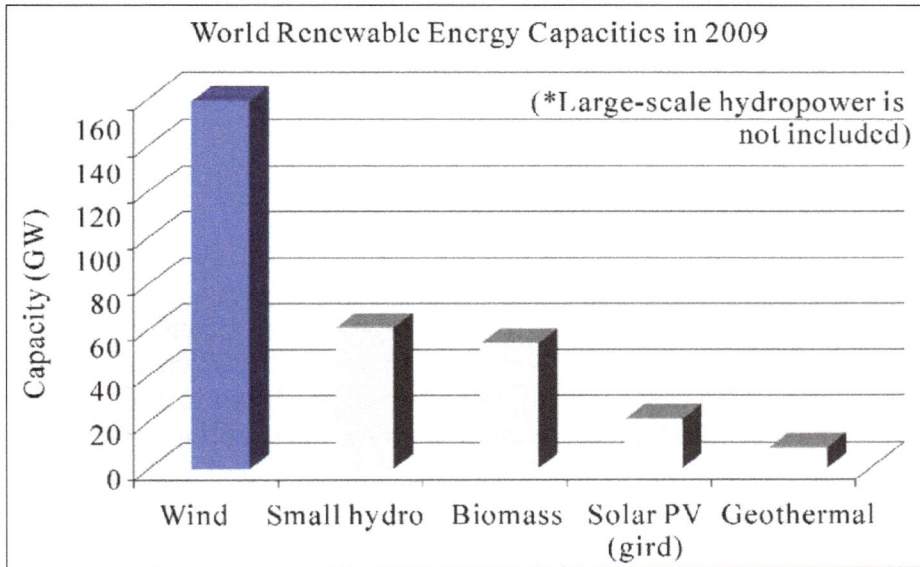

World renewable energy capacities.

There seem small amount of investigation and discussion on some newer concepts of WGTs as well as PECs along with its modulation strategies.

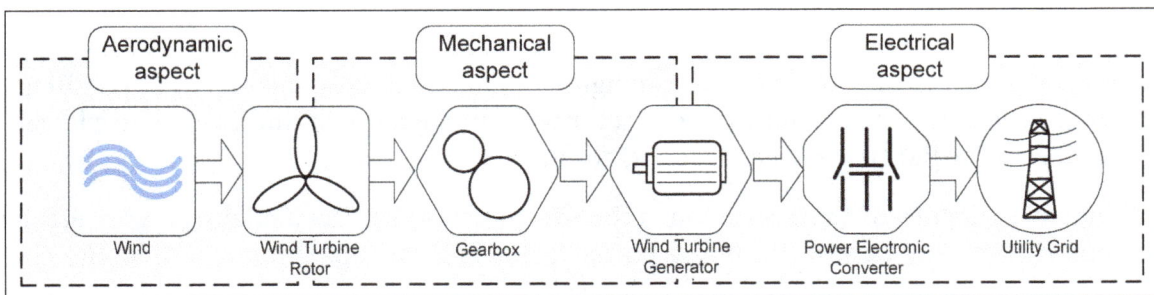

Wind energy conversion system.

## Wind Turbine Generators

## Wind Turbine Generators in the Current Market

WTGs can be classified into three types according to its operation speed and the size of the associated converters as below:

- FSWT (Fixed Speed Wind Turbine).

- VSWT (Variable Speed Wind Turbine) with:

  ◦ PSFC (partial scale frequency converter).

  ◦ FSFC (full scale frequency converter).

FSWT including SCIG (Squirrel-Cage Induction Generator), led the market until 2003 when DFIG (Doubly Fed Induction Generator), which is the main concept of VSWT with PSFC, overtook and has been the leading WTG concept with 85% of the market share reported in 2008. For VSWT with FSFC, WRSG (Wound Rotor Synchronous Generator) has been the main concept; however PMSG (Permanent Magnet Synchronous Generator) has been drawing more attention and increasing its market share in the past recent years due to the benefits of PMSG and drawbacks of WRSG.

## Two Newer WTG Concepts

### BDFIG

BDFIG is one of the most popular VSWT with PSFC types in the current research area due to its inherited characteristics of DFIG, which is the most popular WTG type at the current market, along with its brushless aspect that DFIG do not possess. As shown in figure, BDFIG consists of two cascaded induction machines; one is for the generation and the other is for the control in order to eliminate the use of sliprings and brushes, which are the main drawback of DFIG.

This brushless aspect increases its reliability, which is especially desirable in offshore application. Other advantages are reported in including its capability with low operation speed. On the other hand, BDFIG has relatively complex aspects in its design, assembly and control, which are some of the main disadvantages of BDFIG.

### BDFRG

There is also another brushless and two-cascaded-stator concept of VSWT with PSFC type in the research area, which is BDFRG. As shown in figure, one distinct design compared with BDFIG is its reluctance rotor, which is usually an iron rotor without copper windings, which has lower cost than wound rotor or PM (permanent magnet) rotor.

This design offers some advantages on top of the advantages of BDFIG including higher efficiency, easier construction and control including power factor control capability as well as the cost reduction and higher reliability including its "fail-safe" operating mode due to its reluctance rotor. Due to its very high reliability, reluctance generators have also been of interest in aircraft industry where design challenges such as harsh environment operation and stringent reliability exist. On the other hand, some of the drawbacks for BDFRG exist such as complexity of rotor deign, its larger machine size due to a lower torque-volume ratio and so forth.

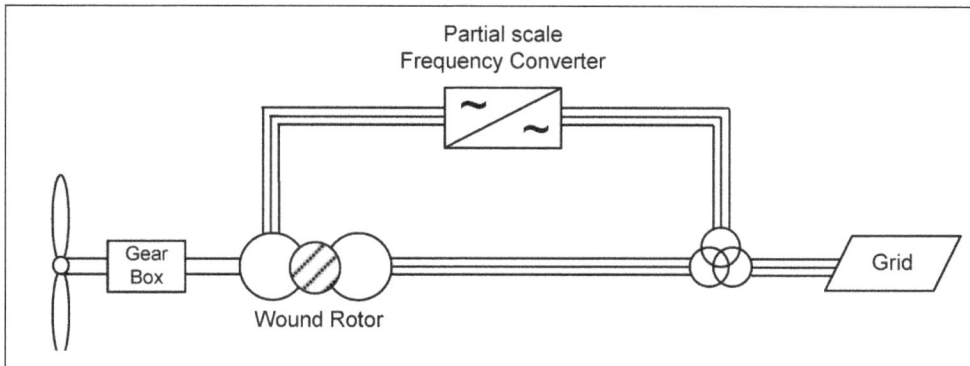

The conceptual diagram of BDFIG.

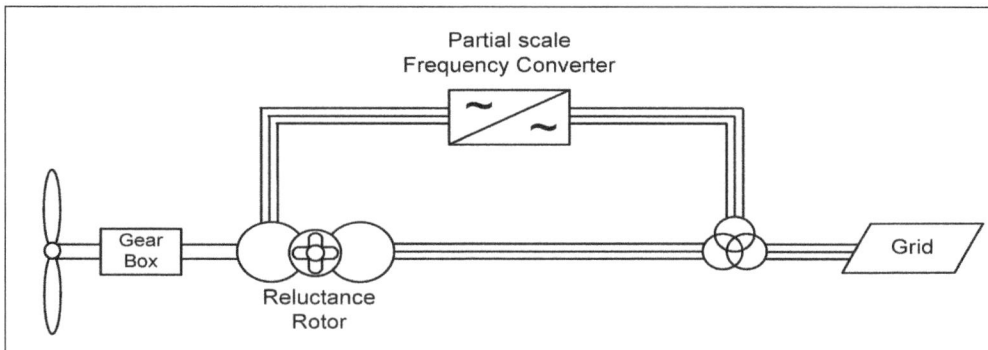

The conceptual diagram of BDFRG.

## Comparison of WTG Concepts

The advantages and disadvantages of the six concepts, the four existing in the current market and the two newer concepts.

Based on the information table represents a comparison of those six concepts with respect to the five criteria; energy yield, cost, reliability, grid support ability and technical maturity. For energy yield, PMSG has the highest rating followed by the other VSWT concepts and SCIG has the lowest energy yield with 10-15% lower value than PMSG due to its fix speed aspect. However, SCIG has the lowest cost followed by BDFRG, and WRSG has the highest cost due to its large size wound machine. It is interesting when 'energy yield per cost' is considered based on the estimated levels on energy yield and cost in table. The highest value is achieved by neither PMSG nor SCIG; BDFRG achieves the highest estimated levels on 'energy yield per cost', which is supported in. Reliability is closely related to the existence of brushes and sliprings, which is the main drawback of DFIG. The reason BDFIG is rated as 'Medium-High' despite of its brushless aspect is because it is new and has design complexity, which brings down reliability as the case of the newer German WTGs compared with older Danish WTGs reported in. On the other hand, BDFRG is rated as 'High' despite of that it is as new concept as BDFIG. It is because of the 'fail-safe' characteristic of BDFRG, which enables its robust operation in spite of the failure on its inverter or secondary stator. Grid support ability is affected mainly by the size of the converter and the stator connection. VSWT with FSFC has high support ability due to its full scale frequency converter. In the case of DFIG, with PSFC, it can only provide limited support to the grid due to its directly connected stator that absorbs the effect of grid fault without any mitigation. It is reported that BDFIG and BDFRG have improved characteristics under grid fault and for grid support ability respectively. Lastly, the maturity of the technology is straightforward as shown in table because SCIG, DFIG and WRSG have been developed for more than a couple of decades followed by PMSG. As mentioned before, BDFIG and BDFRG are newer concept and therefore more researches are needed in order to increase its technical maturity and hence to be applicable in the industry.

## Discussion on WTGs

As observed previously, there has been 'variable speed' trend in the WT market due to its greater energy yield along with other advantages and will be so in the future with DFIG and PMSG leading the market base on the various data and literature The two newer concepts, BDFIG and BDFRG are also in line with this trend.

Another distinct trend is offshore wind energy. It is reported that offshore wind resource has higher quality in terms of its availability and constancy, and higher spatial availability than onshore wind resource, which makes offshore wind very attractive. However, there exist great technical challenges on its construction and maintenance, because of its geological accessibility that greatly depends on the weather condition, which is an unpredictable external factor. Due to this reason, offshore wind has only 1.2% of the world's total installed wind energy share (onshore and offshore) at the current market and is reported to cost 1.5-2 times more than equal-size onshore wind application. As discussed previously, DFIG is less attractive for offshore application due to its pre-planned maintenance for brush and sliprings whereas PMSG, BDFIG and BDFRG are more attractive due to its brushless aspect. BDFRG is especially attractive for its reliability due to its reluctance rotor as discussed previously. Although offshore wind has low level of installation at the present, the growth rate was reported to be 30% in 2009 and is expected to continue to grow.

Lastly, 'Multi-MW' trend is also observed at the current wind turbine market due to the fact that larger power station has lower cost per kWh. The size of the turbine in the current market has gone up to 5-6 MW or even greater, supported by the increased technical level in design and construction. In terms of the cost of the material, DFIG and BDFRG are preferable over PMSG and BDFIG for this trend since PM material in PMSG is costly, and BDFIG has a wound rotor with the two wound cascaded stator, which has greater amount of windings than DFIG or BDFRG.

The advantages and disadvantages of the six WTG concepts:

## SCIG (FSWT)

Advantages:

- Easier to design, construct and control.

- Robust operation.

- Low cost.

Disadvantages:

- Low energy yield.

- No active/reactive power controllability.

- High mechanical stress.

- High losses on gear.

## PMSG (VSWT-FSPC)

Advantages:

- Highest energy yield.

- Higher active/reactive power controllability.

- Absence of brush/slipring.

- Low mechanical stress.

- No copper loss on rotor.

Disadvantages:

- High cost of PM material.

- Demagnetisation of PM.

- Complex construction process.

- Higher cost on PEC.

- Higher losses on PEC.

- Large size.

## WRSG(VSWT-FSPC)

Advantages:

- High energy yield.

- Higher active/reactive power controllability.

- Absence of brush/slipring.

- Low mechanical stress.

Disadvantages

- Higher cost of copper winding.

- Higher cost on PEC.
- Higher losses on PEC.
- Large size.

## DFIG (VSWT-PSPC)

Advantages

- High energy yield.
- High active/reactive power controllability.
- Lower cost on PEC.
- Lower losses by PEC.
- Less mechanical stress.
- Compact size.

Disadvantages:

- Existence of brush/slipring.
- High losses on gear.

## BDFIG (VSWT-PSPC)

Advantages:

- Higher energy yield.
- High active/reactive power controllability.
- Lower cost on PEC.
- Lower losses by PEC.
- Absence of brush/slipring.
- Less mechanical stress.
- Compact size.

Disadvantages:

- Early technical stage.
- Complex controllability, design and assembly.
- High losses on gear.

## BDFRG (VSWT-PSPC)

Advantages:

- Higher energy yield.

- High active/reactive power controllability.

- Lower cost on PEC.

- Lower losses by PEC.

- Absence of brush/slipring.

- No copper loss on rotor.

- Less mechanical stress.

- Easier construction.

Disadvantages:

- Early technical stage.

- Complex controllability and rotor design.

- High losses on gear.

- Larger size than DFIG.

The comparison of the six different WTG concepts.

| Generator Concept | Energy Yield | Cost | Reliability | Grid Support Ability | Technical Maturity |
|---|---|---|---|---|---|
| SCIG | Low | Low | High | Low | High |
| PMSG | High | Medium-High | High | High | Medium-High |
| WRSG | Medium-High | High | High | High | High |
| DFIG | Medium-High | Medium | Medium | Medium | High |
| BDFIG | Medium-High | Medium | Medium-High | Medium-High | Low |
| BDFRG | Medium-High | Low-Medium | High | Medium- High | Low |

## Power Electronic Converters

## Topology of Power Electronic Converters

As the amount of the installed VSWT increased, so has the importance of PECs in WECS since it is the interface between WTGs and the electrical grid. There are three types of converters widely available in the current wind energy market: Back-to-back PWM converter, multilevel converter and matrix converter.

## Back-to-back PWM Converters

Back-to-back PWM converter, which is also referred as 'two-level PWM converter', is the most conventional type among the PEC types for VSWT. As shown in figure, it consists of two PWM-VSIs (voltage source inverters) and a capacitor in between. This capacitor is often referred as a 'DC link capacitor' or 'decoupling capacitor' since it provides a separate control in the inverters on the two sides, which are 'machine' and 'grid' side. In addition, it has lower cost due to its maturity.

However, the DC link capacitor also becomes the main drawback of the PWM converter because it decreases the overall lifetime of the system. There are other disadvantages including switching losses and emission of high frequency harmonics, which results in additional cost in EMI-filters.

## Multilevel Converters

Compared with two-level PWM converter, multilevel (ML) converter has three or more voltage levels, which results in lower total harmonic distortion (THD) than back-to-back PWM converter does. In addition, ML converter offers higher voltage and power capability, which advocates the trend of 'Multi-MW' wind turbine. Another advantage is that switching losses are smaller in ML converter than two-level PWM converter by 25%.

One of the disadvantages on ML converter is the voltage imbalance caused by the DC link capacitors. Another disadvantage in some ML converter designs is uneven current stress on the switches due to its circuit design characteristic. The cost associated with the high more number of switches and the complexity of control are two other drawbacks.

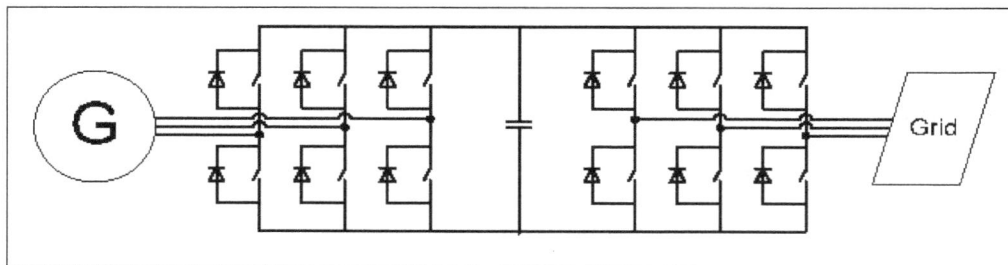

Basic schematics of Back-to-Back PWM converter.

Since the first proposed design of ML converter, the neutral-point clamped three-level converter in 1981, there have been various designs for ML converters including the followings:

- Neutral Point Clamped (NPC) ML converter.

- Cascade Half-Bridge (CHB) ML converter.

- Fly-capacitor (FLC) ML converter.

The detail of each design, which is beyond the scope of this paper, can be found in the literatures.

Out of these three ML converter designs, NPC ML converter is commonly utilised in WECS, especially in multi-MW scale WECS, due to its maturity and advantages. Main drawback exists, however, with 3LNPC (3 level-NPC) design, which is the uneven loss distribution among the semiconductor devices, limiting output power of the converter. This drawback has been overcome with the replacement of the clamping diode with the active switching devices. This modified design of NPC is referred as 'Active NPC' (ANPC), which was first introduced in 2001, as shown in figure. There are many advantages of ANPC including higher power rating than normal NPC by 14% and robustness against the fault condition.

## Matrix Converters

Matrix converters have a distinct difference from the previous two converters in a way that it is an AC-AC converter without any DC conversion in between, which indicates the absence of passive components such as the DC link capacitor and inductor in the converter design. As shown in Figure, the typical design of matrix converters consists of 9 semi-conductors that are controlled with two control rules to protect the converter; three switches in a common output leg must not be turned on at the same time and the connection of all the three output phases must be made to an input phase constantly. There are some advantages of matrix converters. The absence of DC link capacitor results in increased efficiency and overall life time of the converter as well as the reduced size and cost compared with PWM-VSI converter. The thermal characteristic of the matrix converter is also another advantage since it can operate at the temperature up to 300°C, which enables to adopt new technologies such as high temperature silicon carbide devices. On the other hand, some of the reported disadvantages include; the limitation on the output voltage (86% of the input voltage), its sensitivity to the grid disturbances and rapid change of the input voltage, higher conducting losses and higher cost of the switch components than PWM-VSI converter. Further technical details of matrix converter can be found in.

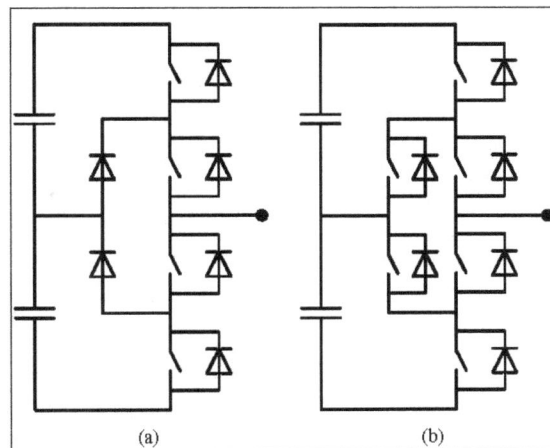

One inverter cell of (a) NPC and (b) ANPC.

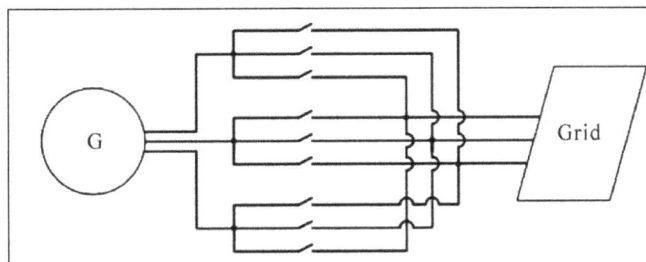

Basic schematics of matrix converter.

## PEC

In terms of power losses, it is widely reported that ML VSCs have less power losses than 2L VSIs with 3-Level Neutral Point Clamped VSIs (3L-NPC) having even lower amount of losses over 3-Level Flying Capacitor VSIs (3L-FLC). This advantage of 3L-NPC, however, inherits poor power loss distribution, which is the main drawback of 3L-NPC as mentioned previously.

Loss distribution is an important aspect in PEC since uneven loss distribution means uneven stress distribution among the semiconductor devices and this results the most stressed switching device to limit the total output power and switching frequency. In uneven loss distribution of 3L-NPC is reported along with other topologies such as 2L-VSI and 3L-FLC, which have even distribution. As mentioned previously, ANPC is the topology to reduce the unevenness among the switching devices and it is reported that 3L-ANPC possess an advantage of 3L-FLC on its natural doubling of switching frequency, without flying-capacitors.

Harmonic performance is another crucial criterion of PEC, especially for WECS as the impact of WECS on power quality of the power grid is increasing due to its increasing penetration level. The comparison on harmonic performance is commonly measured by total harmonic distortion (THD) or weighted THD (WTHD). A comparison on THD of 2L-VSI, 3L-NPC and matrix converter with PMSG is undertaken in and 3L-NPC provides the lowest value of THD among the three topologies. This result verifies that THD decreases with increasing number of levels.

Different PEC topologies consist of components with variable numbers and sizes that result variation in cost. Although 2L-VSI has less number of components compare to ML VSIs, it is estimated to be more costly due to its large LC filter, which is the result of compromise for high efficiency and low THD that ML can achieve with smaller LC filter. Matrix converters would lie in between 2L VSIs and 3L VSIs since it has smaller number of semiconductors and LC filters are required to minimise the switching frequency harmonics. The cost estimation would be similar for both 3L-NPC and 3L-FLC since the excessive cost for the larger LC filter and semiconductors would be compensated with the cost for flying capacitors by considering the cost estimation in with the constant switching frequency. In comparison between 3L-ANPC and 3L-NPC is conducted with different IGBT ratings available in the market. In the literature, it is found that 3L-NPC is most economical (i.e. lest cost per MVA) with 2.3 kV IGBT modules at any switching frequency between 300 Hz to 1050 Hz. However, 3L-ANPC becomes more cost-effective with 3.3 kV and 4.16 kV at switching frequency over 750 Hz.

It is evident that 3L-ANPC is a very attractive PEC topology for WECS, which is increasing its power rating, operates with high switching frequency (typically 2~5 kHz [47,51,54-57]) and requires low harmonic emission.

## Modulation Methods

Along with the converter topologies, there are some modulation strategies available to produce a desired level of output voltage and current in lower frequency. Pulsewidth modulation (PWM) is one of the most widely used modulation strategies for PEC with AC output.

While the primary goal of PWM is to produce a targeted low-frequency output voltage or current, it is also essential for PWM schemes to minimise the impact on the quality of the output signals such as harmonic distortion.

Among the vast amount of proposed PWM schemes, majority of them can be categorised into the following three types despite of different converter topologies.

- Carrier-Based PWM.

- Space Vector Modulation (SVM).

- Selective Harmonic Elimination (SHE).

## Carrier-based PWM

Carrier-based PWM strategy has been widely utilised as the basic logic of generating the switching states is simple. The basic principle is to compare a low frequency sinusoidal reference voltages to high frequency carrier signals, then produce the switching states every time the reference signal intersects carrier signals. The number of carrier signals is defined as (N-1), where N is the number of the level of multi-level VSI (eg. N = 3 for 3-Level NPC VSI).

The basic control diagram and modulation signals of 3-Level VSI are represented in figure.

From the conventional schemes, there are some modified techniques proposed with multi-level or multi-phase methods in order to reduce distortion in ML inverters.

## Space Vector Modulation (SVM)

Space vector modulation (SVM) is the PWM method based on the space vector concept with d-q transformation that is widely utilised in AC machines. With the development of microprocessors, it has become one of the most widely used PWM strategies for three-phase converters due to some of its advantages including high voltage availability, low harmonics, simple digital implementation and wide linear modulation range, which is one of the main aims of PWM.

There are N3 switching states in N-level PWM inverter so in the case of 3-Level NPC VSI, there are 27 (= 3$^3$ ) possible switching states. As shown in figure, these switching states define reference vectors, which are represented by the 19 nodes in the diagram with the four classification of 'zero' ($V_o$), 'small' ($V_{Si}$), 'medium' ($V_{Mi}$) and 'large' ($V_{Li}$), where i = 1,2,…,6. The difference between the numbers of the switching states and space vectors indicate that there is redundancy of switching states existing for some space vectors. As indicated in figure, one 'zero' space vector (i.e. $V_o$) can be generated by three different switching states and six 'small' space vector (i.e. $V_{Si}$) by two different switching states each. These redundancies provide some benefits including balancing the capacitor voltages in 3L-NPC VSI.

The basic principle of SVM is to select three nearest vectors that consist of a triangle in the space vector diagram that the tip of a desired reference vector is located, and generate PWM according to the switching states of those selected vectors. There are many researches on SVM to improve on various aspects such as the improvement in neutral point (NP) balancing at higher modulation indexes and the reduction of the size of DC-link in control loop for renewable application such as WECS.

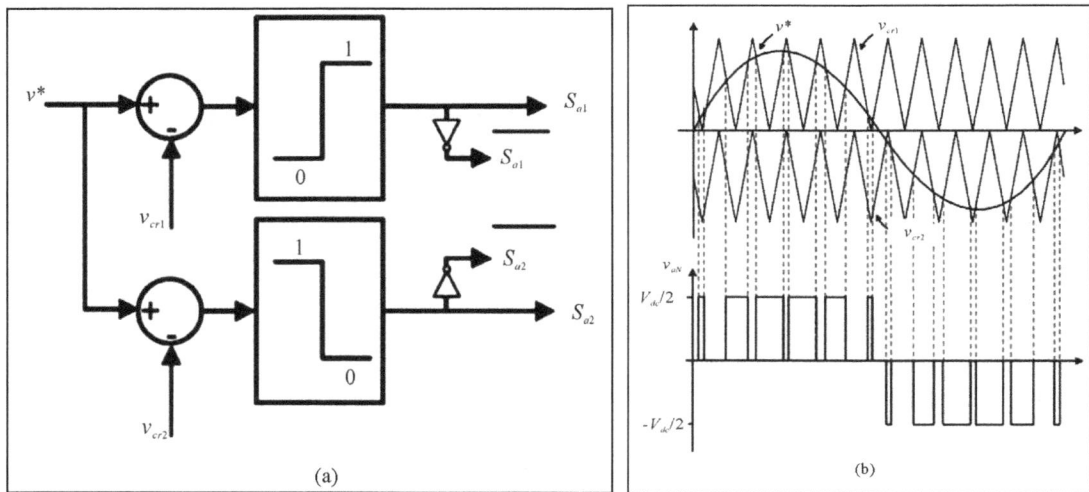

(a) Control diagram, (b) Modulation signal.

## Selective Harmonic Elimination

The basic principle is to calculate N number of switching angles that are less than $\pi/2$ for a N-Level inverter through N number of the nonlinear equation with Fourier expansion of output voltage. One equation is used to control the fundamental frequency through the modulation index and the other N-1 equations are used for elimination of the low-order harmonics components. In the case of 3-Level VSI, $5^{th}$ and $7^{th}$ harmonic components are the two lowest-order harmonics to be eliminated since $3^{rd}$ harmonic component is cancelled by the nature of three-phase. Figure depicts an example of the 3-Level SHE with 3 switching angles, $a_1$, $a_2$ & $a_3$.

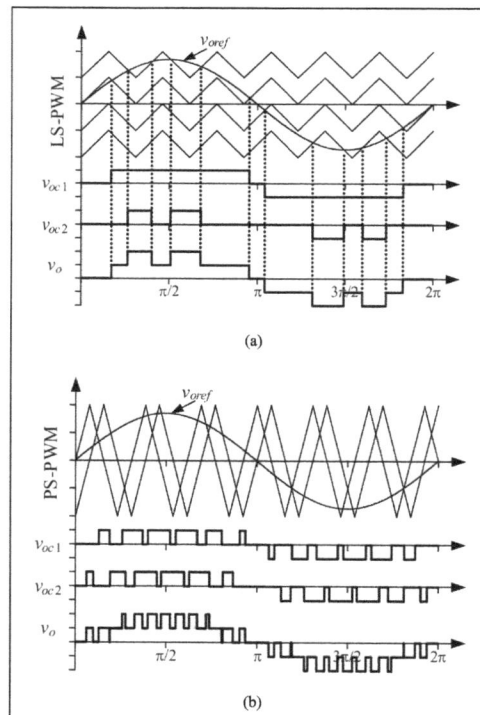

(a) Multi-level PWM, (b) Multi-phase PWM.

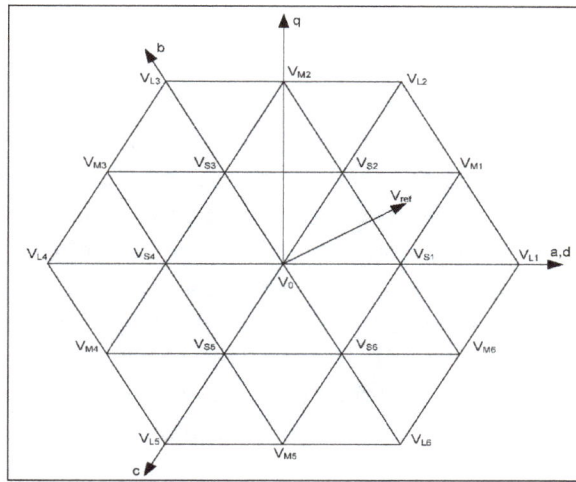

Normalised space vector diagram for the threelevel NPC converter.

It is well-known that SHE strategy provides good harmonic performance in spite of the low switching frequency due to its harmonic elimination nature. Another advantage is the reduction on its switching loss due to the low switching frequency. However, there are some disadvantages exist including its heavy computational cost and narrow modulation range. There are many researches on SHE such as NP balancing for 3L-NPC and the increase of the number of eliminating low-order harmonics with simple in formulation.

## Modulation Method

Among the three modulation methods discussed above, CB-PWM and SVM are widely utilised in WECS. However, SHE strategy has not been utilised in WECS to the best knowledge of the author despite of its active researches with resent PEC technologies such as 5-Level ANPC VSI. The authors in suggest the combination of using 2L SVM and SHE schemes for the switching frequency fsw ≤ 500 Hz whereas the combination of 2L SVM and 3L SVM for fsw > 500 Hz due to their performances with respect to the modulation index and switching frequency. This could be one reason for SHE schemes not to be utilised in WECS where high switching frequency is used.

However, if the reason of high switching frequency in WECS is for the quality of output power, lower switching frequency can be adapted with SHE strategy in WECS for high quality of output power. This would increase the efficiency of WECS due to less switching losses and also this will reduce a cost of filter circuits since the size of the filter would be smaller with the nature of harmonic elimination of SHE.

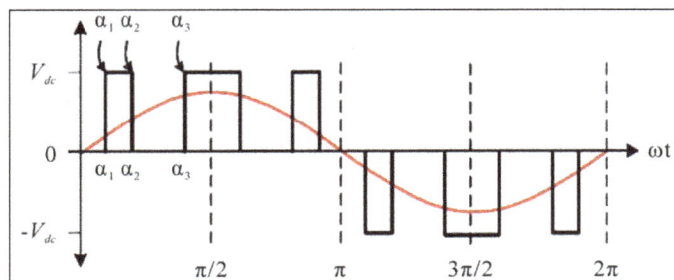

3-level SHE.

## Issues on Grid-connection

### The Utility Grid and WECS

In the utility grid, some grid disturbances such as voltage dips often occur. In the past, grid-connected wind turbines needed to be disconnected from the grid when such disturbances happened in order to protect themselves from damages. However, as the penetration level of wind energy has been increasing, especially in the last decade, the role of WECS on the grid has been transforming from minor power source to main power supply stations such as coal-fired power stations along with the new grid codes. Fault Ride Through (FRT) capability under voltage dip is one of the main focus of the new grid codes that came into effect by the German utility company, E.ON, in Germany in 2004 and in other countries. Another focus of the new grid code is the requirement for wind turbines to support the power quality control on the grid such as voltage/frequency stability control, active/reactive power regulation, harmonics/inter-harmonics emission and flicker emission and so forth. International Electrotechnical Commission (IEC) has also released standards on power quality for grid-connected wind turbines, which are IEC 61400-21 in 2001 and its second edition in 2008.

### The Three Main Issues in Grid-connected WECS

### Voltage Dip

Voltage dip, also referred as voltage sag, is a phenomenon that the voltage of the grid drops below the normal rms level (down to 0.1-0.9 p.u.) for a short duration (typically 0.5-30 cycles) [81,82]. It is a critical issue for wind turbines because the voltage dip can initiate abnormal behaviours in the generator and PEC, which can result in permanent damages. Therefore it is regarded as a significant technical challenge for wind turbine manufacturers. Under the new grid codes, wind turbines are expected to have reasonable FRT capability, which is to support the grid under the voltage dip as well as to protect themselves from being damaged. Figure represents a typical FRT capability curve by E.ON. Wind turbine must stay connected until the state (i.e. voltage-time) is placed below the solid line in the figure in order to support the grid. There have been various attempts on FRT capability including crowbar protection, GSC and MSC controllability, and so forth, and further details on FRT capability can be found in the literatures.

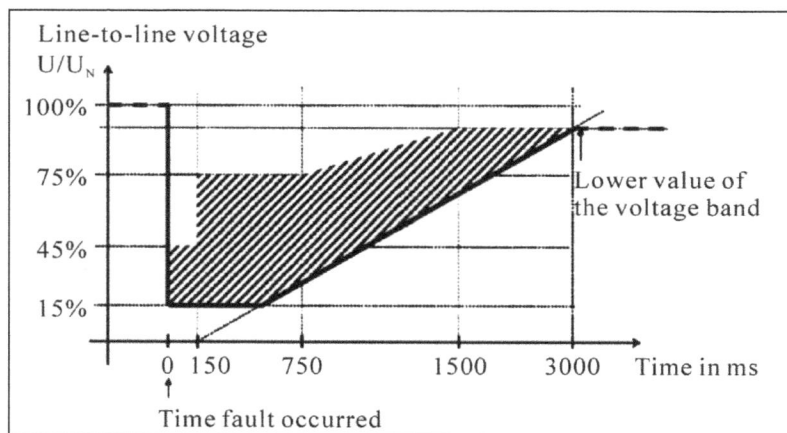

The FRT capability curve.

## Harmonic Emission

Harmonic emission is another crucial issue for gridconnected wind turbines because it may result in voltage distortion and torque pulsations, which consequently causes overheating in the generator and other problems. Although wind turbines emit low-order harmonics by nature, self-commutated converters used in modern VSWTs can filter out this low-order harmonics. However, these self-commutated converters introduce high-order harmonics instead. In addition, interharmonics, which is non-integer harmonics, is another type of harmonic emission by WTCSs. It contributes to the level of the flicker and has an interference with control and protection signals in power lines, which are regarded as the most harmful effects on the power system.

Wind turbine power quality standard IEC 61400-21 2nd edition released in 2008, along with harmonic measurement standard IEC 61000-4-7, provides the requirements for on current harmonics, current interharmonics and higher current components to be measured and reported in modern WECSs.

## Flicker

Flicker is another issue on wind turbine associated with the grid. Flicker is defined as a measure of annoyance of flickering light bulbs on human, caused by active and reactive power fluctuation as a result of the rapid change in wind speed. The standard IEC 61400-21 requires flicker to be monitored in two operation modes; continuous operation and switching operation. The switching operation is the condition of cut-in and cut-out by wind turbine. It is reported that flicker is relatively less critical issue in VSWT; however, it needs to be improved for higher power quality.

The onshore wind is the majority with the share of 98.8% of the current wind turbine market, in which 85% is utilising DFIG concept. However, offshore wind has been gaining more and more attention due to its rich wind resource and hence, more researches are intensively being undertaken on offshore wind. Therefore, future wind turbine market is expected to have more number of offshore wind turbines with the brushless design such as PMSG, BDFIG and especially BDFRG, due to its high reliability as discussed previously. 'Multi-MW' trend is also observed and BDFRG seems to have favourable characteristic for this trend among the brushless WTG designs because of its reluctance rotor. The trend also affects on the design of PECs, resulting in the preference of ML converter, especially ANPC, due to its higher voltage capability, reduced switching losses and its cost-effectiveness with IGBT modules with higher voltage rating. The increased concerns on harmonics, which is one of the discussed grid-connection issues, also make ML converter more attractive than other PEC topologies due to its lower harmonics emission. BDFRG is also reported to have lower harmonic emission to the grid, making this technology greatly suitable to meet the demand of the current and the future wind energy market.

# Wind Power Calculations

## 4

- **Calculation of Wind Power**
- **Wind Turbine Blade Efficiency and Power Calculation**
- **Power Output of Wind**
- **Physics of Wind Turbines**

The power output of the wind and wind turbine blade efficiency can be calculated by using Betz law, kinetic energy and potential energy equations. This chapter closely examines these laws and equations associated with wind power to provide an extensive understanding of the subject.

## Calculation of Wind Power

There are many complicated calculations and equations involved in understanding and constructing wind turbine generators however the layman need not worry about most of these and should instead ensure they remember the following vital information:

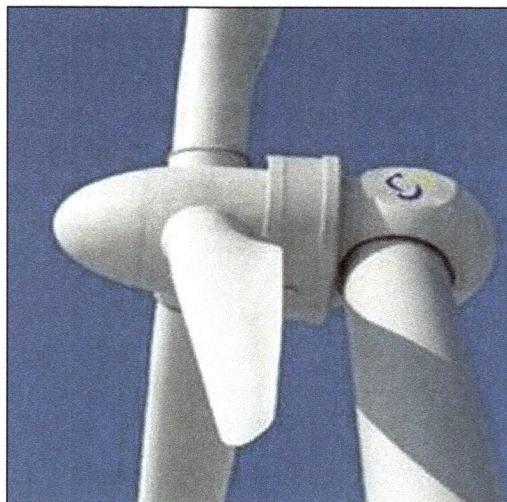

- The power output of a wind generator is proportional to the area swept by the rotor – i.e. double theswept area and the power output will also double.

- The power output of a wind generator is proportional to the cube of the wind speed – i.e. double the wind speed and the power output will increase by a factor of eight (2 x 2 x 2).

## The Power of Wind

Wind is made up of moving air molecules which have mass – though not a lot. Any moving object with mass carries kinetic energy in an amount which is given by the equation:

Kinetic Energy = 0.5 x Mass x Velocity$^2$

where the mass is measured in kg, the velocity in m/s, and the energy is given in joules.

Air has a known density (around 1.23 kg/m$^3$ at sea level), so the mass of air hitting our wind turbine (which sweeps a known area) each second is given by the following equation:

Mass/sec (kg/s) = Velocity (m/s) x Area (m$^2$) x Density (kg/m$^3$)

And therefore, the power (i.e. energy per second) in the wind hitting a wind turbine with a certain swept area is given by simply inserting the mass per second calculation into the standard kinetic energy equation given above resulting in the following vital equation:

Power = 0.5 x Swept Area x Air Density x Velocity$^3$

where Power is given in Watts (i.e. joules/second), the Swept area in square metres, the Air density in kilograms per cubic metre, and the Velocity in metres per second.

## Wind Turbine Blade Efficiency and Power Calculation

Power production from a wind turbine is a function of wind speed. The relationship between wind speed and power is defined by a power curve, which is unique to each turbine model and, in some cases, unique to site-specific settings. In general, most wind turbines begin to produce power at wind speeds of about 4 m/s (9 mph), achieve rated power at approximately 13 m/s (29 mph), and stop power production at 25 m/s (56 mph). Variability in the wind resource results in the turbine operating at continually changing power levels. At good wind energy sites, this variability results in the turbine operating at approximately 35% of its total possible capacity when averaged over a year.

The amount of electricity produced from a wind turbine depends on three factors:

- Wind speed: The power available from the wind is a function of the cube of the wind speed. Therefore if the wind blows at twice the speed, its energy content will increase eight-fold. Turbines at a site where the wind speed averages 8 m/s produce around 75-100% more electricity than those where the average wind speed is 6 m/s.

- Wind turbine availability: This is the capability to operate when the wind is blowing, i.e. when the wind turbine is not undergoing maintenance. This is typically 98% or above for modern European machines.

- The way wind turbines are arranged: Wind farms are laid out so that one turbine does not take the wind away from another. However other factors such as environmental considerations, visibility and grid connection requirements often take precedence over the optimum wind capture layout.

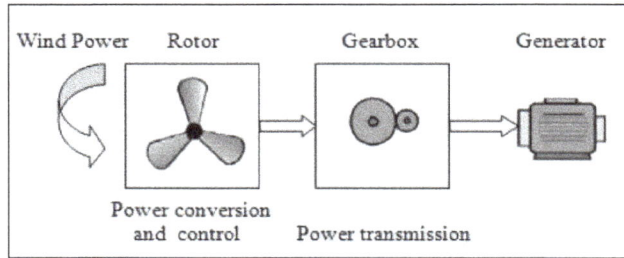

Mechanical components of wind turbine.

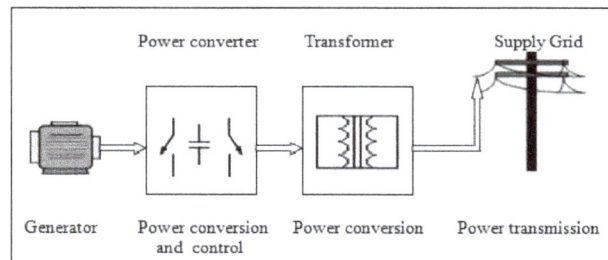

Electrical components of wind turbine.

Usually, WTBs are designed to operate for a period of 20 years. But, no final statement can be made yet concerning the actual life expectancy of modern WTBs as, until now, no operational experience of such period is available.

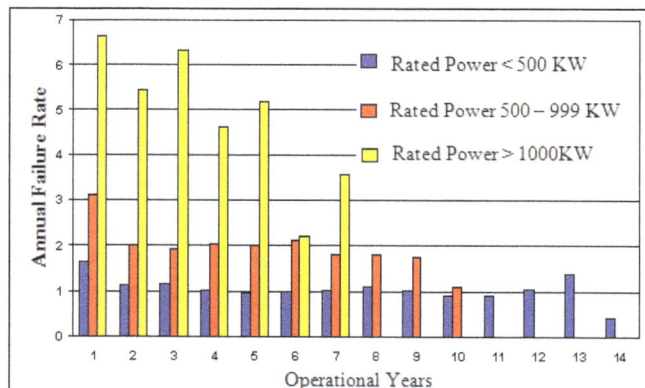

Frequency of 'failure rate' with increasing operational age.

Changes in reliability with increasing operational age can, however, provide indications of the expected lifetime and the amount of upkeep required. Reliability can be expressed by the number of failures per unit of time, i. e. 'Failure Rate'. In the following, the failure rates of WTBs depending on their operational age will be depicted.

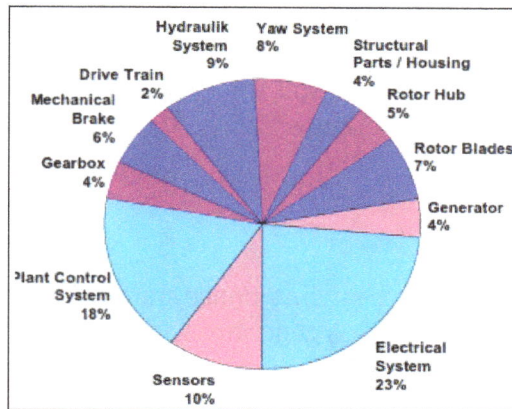

Share of the main components of total number of failures.

It is clear that the failure rates of the WTs now installed, have almost continually declined in the first operational years. This is true for the older turbines under 500 kW and for the 500/600 kW class. However, the group of mega-watt WTs show a significantly higher failure rate, which also declines by increasing age. But, including now more and more mega-watt. WTB models of the newest generation, the failure rate in the first year of operation is being reduced.

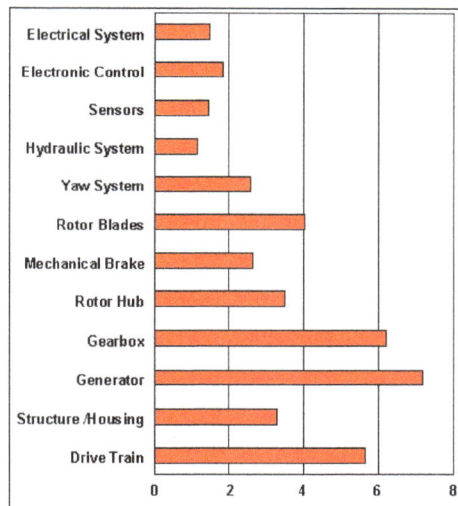

Down time of wind turbine system components.

Wind turbines achieve an excellent technical availability of about 98% on average, although they have to face a high number of malfunctions.

## Improving System Liability

## Identify Critical Components

Within any complex system, certain components will stand out as high-risk items, either because they are 'weak points' that are demonstrated to be failure prone, are absolutely essential to turbine operation, or are expensive and time consuming to diagnose and repair. Identifying the critical components allows the O&M staff to direct their monitoring, training, inventory, and logistics efforts on areas that will provide the most benefit.

Although to some extent the critical components depend on the manufacturer, configuration, and operating environment, certain candidates for attention (gearboxes, generators, and power converters, for example) are well known throughout the industry. Minor components, though perhaps less costly to replace or repair, may be elevated to a critical status if their frequency of failure is high.

## Characterize Failure Modes

Understanding the failure mode allows the maintenance staff to focus monitoring efforts and potentially delay or prevent catastrophic failures. A generator short may be difficult to predict, but gearbox bearing or gear wear may be detected early with scrupulous lubricant monitoring and "condition" monitoring, and the progression of damage possibly mitigated with more frequent oil changes or better filtering. An understanding of the way in which a failure progresses is essential to ensuring that staff avoid consequential damage due to unanticipated breakage.

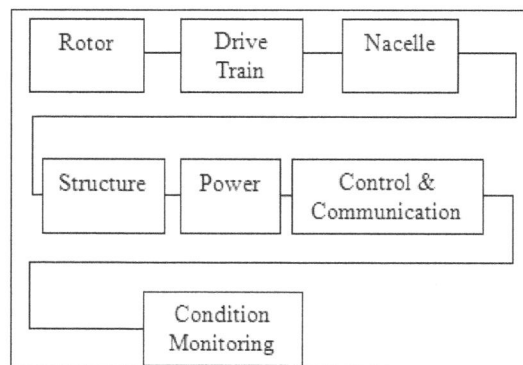

Reliability block diagram of wind turbine.

## Determine the Root Cause

Although the wind plant operator may be primarily interested in replacing a failed component and getting their machine back on-line, a failure always represents an opportunity for improvement. Most wind turbine manufacturers include failure analysis as an essential part of their continuous quality improvement process. Evaluating the root cause of a major component failure is essential to determining if the failure is due to manufacturing quality, product misapplication, design error, or inappropriate design assumptions. This information, in turn, assists the manufacturer in determining if the problem is an isolated instance or a systemic problem that is likely to result in serial failures. In the latter case, retrofits or redesigns will be required and a field replacement plan will be developed.

It can be assumed that these good availability figures can only be achieved by a high number of service teams who respond to turbine failures within short time. In order to further improve the reliability of WTBs, the designers have to better the electric and electronic components. This is particularly true and absolutely necessary in the case of new and large turbines.

## Turbine Blade Design

Tip Seep Ratio(TSR): Select a value for the Tip Speed Ratio (TSR) which is defined as: TIP SPEED

RATIO (TSR) = (tip speed of lade)/(wind speed). The tip speed ratio is a very important factor in the different formulas of blade design. Generally can be said, that slow running multi bladed wind turbine rotors operate with tip speed ratios like 1-4, while fast runners use 5-7 as tip speed ratios.

The task is now to fit the known generator capacity and revolutions to the wind speed and to the swept rotor area. Two formulas are needed.

Power (W) = 0.6 x Cp x N x A x $V^3$, Revolutions (rpm) = V x TSR x 60 / (6.28 x R), Cp = Rotor efficiency, N = Efficiency of driven machinery, A = Swept rotor area (m2 ), V = Wind speed (m/s) TSR = Tip Speed Ratio, R = Radius of rotor, Rotor efficiency can go as high as Cp = 0.48, but Cp = 0.4 is often used in this type of calculations.

This concept works without transmission. If a transmission with efficiency of 0.95 was to be included this means that N = 0.95 x 0.7, Tip speed ratio "TSR" = 7" Wind speed "V" = 8.6 m/s Rotor efficiency "Cp" = 0.4 Generator efficiency "N" = 0.7 Swept rotor area "A" = 2.11 M2 Radius of rotor = 0.82 m Revolutions = 701 rpm Power output = 226 W.

The width of the blade is also called the blade chord. A good formula for computing this is: Blade Chord (m) = 5.6 x R2 / (i x Cl x r x $TSR^2$ ), R = Radius at tip, r = radius at point of computation, i = number of blades, Cl = Lift coefficient, TSR = Tip Speed Ratio.

## Calculation of Wind Power

There are many complicated calculations and equations involved in understanding and constructing wind turbine generators however the layman need not worry about most of these and should instead ensure they remember the following vital information:

- The power output of a wind generator is proportional to the area swept by the rotor - i.e. double the swept area and the power output will also double.

- The power output of a wind generator is proportional to the cube of the wind speed.

Kinetic Energy = 0.5 x Mass x $Velocity^2$, where the mass is measured in kg, the velocity in m/s, and the energy is given in joules. Air has a known density (around 1.23 kg/$m^3$ at sea level), so the mass of air hitting our wind turbine. (which sweeps a known area) each second is given by the following equation: Mass/sec (kg/s) = Velocity (m/s) x Area ($m^2$ ) x Density (kg/$m^3$ ). Therefore, the power (i.e. energy per second) in the wind hitting a wind turbine with a certain swept area is given by simply inserting the mass per second calculation into the standard kinetic energy equation given above resulting in the following vital equation: Power = 0.5 x Swept Area x Air Density x $Velocity^3$, where Power is given in Watts (i.e. joules/second), the Swept area in square metres, the Air density in kilograms per cubic metre, and the Velocity in metres per second.

The equation for wind power (P) is given by P= 0.5 x ρ x A x Cp x $V^3$ x Ng x Nb where, ρ = Air density in kg/$m^3$, A = Rotor swept area ($m^2$). Cp = Coefficient of performance V = wind velocity (m/s) Ng = generator efficiency Nb = gear box bearing efficiency.

The world's largest wind turbine generator has a rotor blade diameter of 126 metres and so the rotors sweep an area of PI x $(diameter/2)^2$ = 12470 $m^2$ ! As this is an offshore wind turbine, we know it is situated at sea-level and so we know the air density is 1.23 kg/$m^3$. The turbine is rated at 5MW in 30mph (14m/s) winds, and so putting in the known values will give, Wind Power = 0.5 x 12,470 x 1.23 x (14 x 14 x 14), which gives us a wind power of around 21,000,000 Watts. Why is the power of the wind (21MW) so much larger than the rated power of the turbine generator (5MW)? Because of the Betz Limit, and inefficiencies in the system. The Betz law means that wind turbines can never be better than 59.3% efficient. The law can be simply explained by considering that if all of the energy coming from wind movement into the turbine were converted into useful energy then the wind speed afterwards would be zero. But, if the wind stopped moving at the exit of the turbine, then no more fresh wind could get in - it would be blocked. In order to keep the wind moving through the turbine, to keep getting energy, there has to be some wind movement on the outside with energy left in it. There must be a 'sweet spot' somewhere and there is, the Betz limit at 59.3%.

## Wind Turbine Blade Calculation and Power Calculation

Power, rotational speed and torque of the wind turbine can be calculated using the free software Blade Calculator.

Table: Output obtained for the given parameters.

| Power | 1.02 KW |
| Rotational speed | 77 rad/sec |
| Torque | 13.25 Nm |

The program display three power. The first power is the power available in the wind for given wind speed and radius. The second power is the Betz Power. This is the maximum power can be extracted from the wind if there were no losses or inefficiencies in the system. Please note that it is very difficult to get the Betz power from any turbine. The last power which is labeled as Real Power is the power you can get from the turbine with the efficiency displayed in the screen. This kind turbines will have 0 efficiency for wind speeds above 25 m/s, meaning that turbine reached it is cut of speed and turned of to prevent damage to turbine.

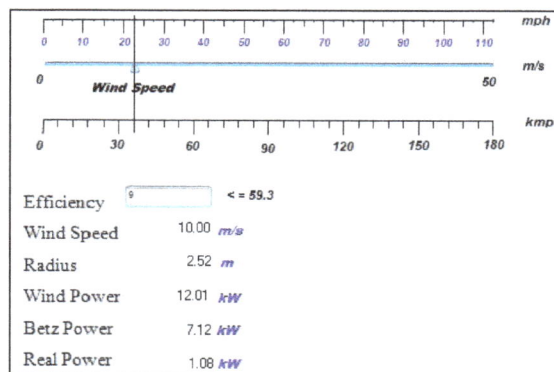

Result obtained for a wind speed of 10m/s and turbine efficiency 9%.

## Calculation of Generator Efficiency

The following conditions/status are assumed. 1. The 3 phases are isolated, and connected as 3 single phase outputs.

Each output is rectified to DC using asingle phase bridge rectifier.

At 666rpm, generator voltage Vs = 65 Volts. $R_s$ = Resistance of each phase of the generator (5.6 Ohms), Voltage across Rs = 65 - 48 = Vs = 17 Volts, V = IR and therefore V/R = I, Current into battery = 17/5.6 = 3 amps per phase, P = VI. Power into battery = 48 x 3 = 144 watts per phase (432 watts for all 3 phases), P = $V^2$/R, Power Lost is = $17^2$/5.6 = 51.6 per phase, Efficiency of generator = 144/(144+51.6) = 73.6% (Theoretical)

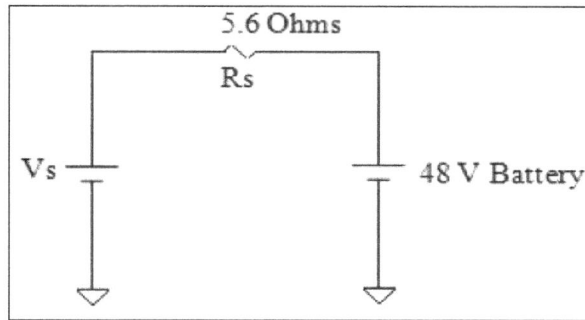

One phase representation of generator.

## Measured Results

$R_s$ is the resistance of the generator windings plus the power cable; 5.75 ohms and $R_1$ is the resistance of the load; 6.6, 10, 15, 21.5 an 25 ohms.

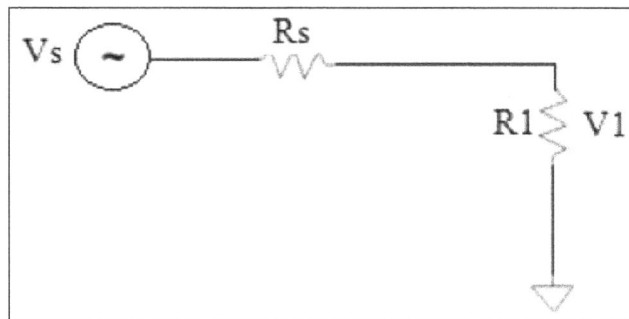

Electrical analogy of wind turbine.

Table: Specification and result for 1kw generator.

| Parameter | Unit | Value |
|---|---|---|
| Rated power | W | 1000 |
| Rated speed | rpm | 300 |
| Rated frequency | Hz | 40 |
| Rated EMF (per coil) | V | 33.6 |
| Number of phases | - | 3 |
| Number of pole pairs | - | 8 |
| Number of armature coils | - | 12 |
| Generator diameter | mm | 462 |
| Generator length | mm | 55 |

Table: Rotational speed of turbine for different wind speed(rpm).

| Wind Speed | 25 ohms | 21.5 ohms | 15 ohms | 10 ohms | 6 ohms |
|---|---|---|---|---|---|
| 30km/h | 820 | 766 | 809 | - | - |
| 40km/h | 1302 | 1363 | 851 | 645 | - |
| 50km/h | 1753 | 1676 | 1489 | 1291 | 1105 |
| 60km/h | - | 2365 | 2098 | 1744 | 1607 |

Table: Power in Watts.

| Wind Speed | 25 ohms | 21.5 ohms | 15 ohms | 10 ohms | 6 ohms |
|---|---|---|---|---|---|
| 30km/h | 208 | 205 | 300 | - | - |
| 40km/h | 524 | 649 | 332 | 252 | - |
| 50km/h | 950 | 981 | 1017 | 1008 | 940 |
| 60km/h | - | 1953 | 2019 | 1837 | 1990 |

Table: Blade efficiency of turbine.

| Wind Speed | 25 ohms | 21.5 ohms | 15 ohms | 10 ohms | 6 ohms |
|---|---|---|---|---|---|
| 30km/h | 0.23 | 0.23 | - | - | - |
| 40km/h | 0.24 | 0.30 | 0.15 | - | - |
| 50km/h | 0.22 | 0.23 | 0.24 | 0.24 | - |
| 60km/h | - | 0.27 | 0.27 | 0.25 | 0.27 |

Table: Tip speed (km/hour).

| Wind Speed | 25 ohms | 21.5 ohms | 15 ohms | 10 ohms | 6 ohms |
|---|---|---|---|---|---|
| 30km/h | 278 | 260 | 275 | - | - |
| 40km/h | 441 | 463 | 289 | 218 | - |
| 50km/h | 595 | 569 | 506 | 438 | 375 |
| 60km/h | - | 803 | 712 | 592 | 546 |

Table: Tip speed ratio.

| Wind Speed | 25 ohms | 21.5 ohms | 15 ohms | 10 ohms | 6 ohms |
|---|---|---|---|---|---|
| 30km/h | 9.2 | 8.7 | 9.2 | - | - |
| 40km/h | 11.0 | 11.6 | 7.2 | 5.5 | - |
| 50km/h | 11.9 | 11.4 | 10.1 | 8.8 | 7.5 |
| 60km/h | - | 16.1 | 14.2 | 11.8 | 10.9 |

Power generated by the blades was calculated by dividing by the efficiency of the generator. Once the blades have been characterized, a new generator will be designed. Powers generated by the blades are calculated by the following:

Voltage across the resistor load was measured $V_1$,

$V_s = V_1 \times [(R_s + R_l) / R_l]$. Power produced by blades, and lost in generator, power cable and resistor load is given by; $P = V_2 / R$ $P = V_s^2 / (R_s + R_l)$

A 1kW @ 11m/s, 1 meter diameter wind turbine designed with the support of software. The wind turbine blades power and efficiency has been measured at different tip-speed-ratios as well as calculated using software tool. The wind turbine blades power and efficiency has been measured at different tip-speed-ratios and a maximum efficiency of 30% at a TSR of 11.6 was recorded, verifying the blade calculator's accuracy. The environmental factors like wind direction, corrosion, water vapour intrusion, thermal expansion, mechanical load, summer-winter climate change, ageing and component derating which degrades the performance can be considered for estimating a more practical and accurate design.

## Power Output of Wind

The following formula illustrates factors that are important to the performance of a wind turbine. Notice that the wind speed, V, has an exponent of 3 applied to it. This means that even a small increase in wind speed results in a large increase in power. Read How high should your small wind turbine be for more information. That is why a taller tower will increase the productivity of any wind turbine by giving it access to higher wind speeds as shown in the Wind Speeds Increase with Height graph. The formula for how to calculate power is:

$$Power = k\, C_p\, 1/2 \rho\, AV^3$$

Where,

P = Power output, kilowatts.

$C_p$ = Maximum power coefficient, ranging from 0.25 to 0.45, dimension less (theoretical maximum = 0.59).

ρ = Air density, lb/ft3.

A = Rotor swept area, ft2 or π D2/4 (D is the rotor diameter in ft, π = 3.1416).

V = Wind speed, mph.

k = 0.000133 A constant to yield power in kilowatts. (Multiplying the above kilowatt answer by 1.340 converts it to horse- power i.e. 1 kW = 1.340 horsepower).

The rotor swept area, A, is important because the rotor is the part of the turbine that captures the wind energy. So, the larger the rotor, the more energy it can capture.

The air density, ρ, changes slightly with air temperature and with elevation. The ratings for wind turbines are based on standard conditions of 59° F (15° C) at sea level. A density correction should be made for higher elevations as shown in the Air Density Change with Elevation graph. A correction for temperature is typically not needed for predicting the long-term performance of a wind turbine.

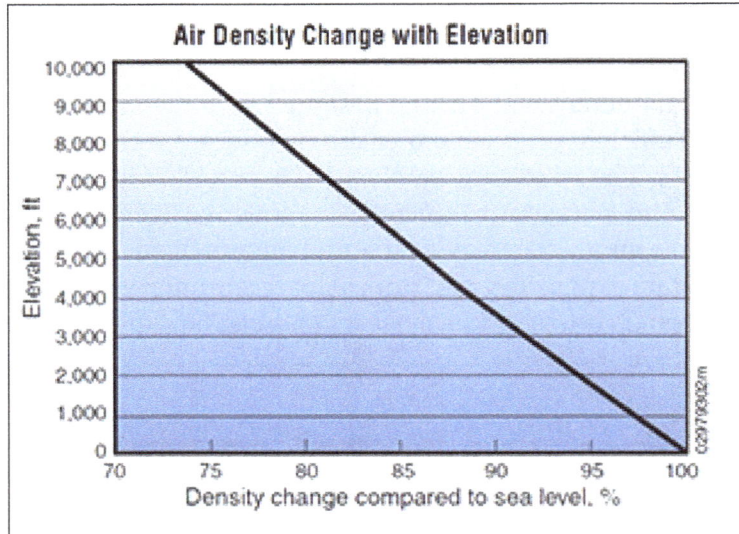

Although the calculation of wind power illustrates important features about wind turbines, the best measure of wind turbine performance is annual energy output. The difference between power and energy is that power (kilowatts [kW]) is the rate at which electricity is consumed, while energy (kilowatt-hours [kWh]) is the quantity consumed. An estimate of the annual energy output from your wind turbine, kWh/year, is the best way to determine whether a particular wind turbine and tower will produce enough electricity to meet your needs.

A wind turbine manufacturer can help you estimate the energy production you can expect. They will use a calculation based on the particular wind turbine power curve, the average annual wind speed at your site, the height of the tower that you plan to use, and the frequency distribution of the wind—an estimate of the number of hours that the wind will blow at each speed during an average year. They should also adjust this calculation for the elevation of your site. Contact a wind turbine manufacturer or dealer for assistance with this calculation.

To get a preliminary estimate of the performance of a particular wind turbine, use the formula below.

$$AEO = 0.01328 \, D_2 \, V_3$$

Where,

AEO = Annual energy output, kWh/year.

D = Rotor diameter, feet.

V = Annual average wind speed, mph.

# Physics of Wind Turbines

All wind turbines, wind power is proportional to wind speed cubed. Wind energy is the kinetic energy of the moving air. The kinetic energy of a mass m with the velocity v is:

$$E_{kin} = \tfrac{1}{2} m v^2$$

The air mass m can be determined from the air density ρ and the air volume V according to:

$$m = \rho V$$

Then,

$$E_{kin,wind} = \tfrac{1}{2} V \rho v^2$$

Power is energy divided by time. We consider a small time, Δt, in which the air particles travel a distance s = v Δt to flow through. We multiply the distance with the rotor area of the wind turbine, A, resulting in a volume of:

$$\Delta V = A v \Delta t$$

which drives the wind turbine for the small period of time. Then the wind power is given as:

$$P_{wind} = \frac{E_{kin,wind}}{\Delta t} = \frac{\Delta V \rho v^2}{2 \Delta t} = \frac{\rho A v^2}{2}$$

The wind power increases with the cube of the wind speed. In other words: doubling the wind speed gives eight times the wind power. Therefore, the selection of a "windy" location is very important for a wind turbine.

The effective usable wind power is less than indicated by the above equation. The wind speed behind the wind turbine can not be zero, since no air could follow. Therefore, only a part of the kinetic energy can be extracted. Consider the following picture.

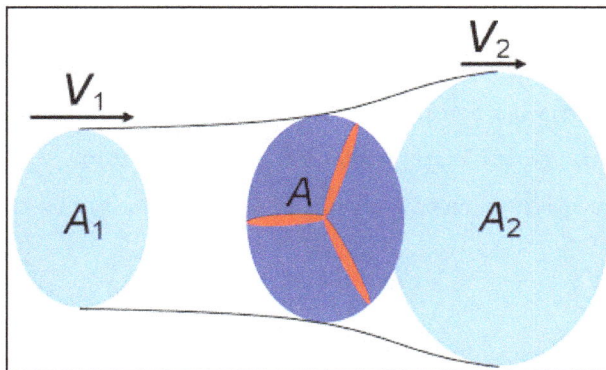

The wind speed before the wind turbine is larger than after. Because the mass flow must be continuous, the area $A_2$ after the wind turbine is bigger than the area $A_1$ before. The effective power is the difference between the two wind powers:

$$P_{eff} = P_1 - P_2 = \frac{\Delta V \rho}{2\Delta t}\left(v_1^2 - v_2^2\right) = \frac{\rho A}{4}\left(v_1 + v_2\right)\left(v_1^2 - v_2^2\right)$$

If the difference of both speeds is zero, we have no net efficiency. If the difference is too big, the air flow through the rotor is hindered too much. The power coefficient $c_p$ characterizes the relative drawing power:

$$C_p = \frac{P_{eff}}{P_{wind}} = \frac{\left(v_1 + v_2\right)\left(v_1^2 - v_2^2\right)}{2v_1^3} = \frac{(1+x)\left(1-x^2\right)}{2}$$

To derive the above equation, the following was assumed: $A_1 v_1 = A_2 v_2 = A (v_1 + v_2) / 2$. We designate the ratio $v_2/v_1$ on the right side of the equation with x. To find the value of x that gives the maximum value of $C_p$, we take the derivative with respect to x and set it to zero. This gives a maximum when x = 1/3. Maximum drawing power is then obtained for $v_2 = v_1 / 3$, and the ideal power coefficient is given by:

$$C_p = \frac{P_{eff}}{P_{wind}} = \frac{16}{27} \approx 59\%$$

Another wind turbine located too close behind would be driven only by slower air. Therefore, wind farms in the prevailing wind direction need a minimum distance of eight times the rotor diameter. The usual diameter of wind turbines is 50 m with an installed capacity of 1 MW and 126 m with a 5-MW wind turbine. The latter is mainly used offshore.

The installed capacity or rated power of a wind turbine corresponds to an electrical power output of a speed between 12 and 16 m/s, with optimal wind conditions. For safety reasons, the plant does not produce greater power at the high wind conditions than those for which it is designed. During storms, the plant is switched off. Throughout the year, a workload of 23% can be reached inland. This increases to 28% on the coast and 43% offshore.

## References

- Calculate-wind-power-output, construction: windpowerengineering.com, Retrieved 23 June, 2019
- Calculation-of-wind-power, wind, wordpress: reuk.co.uk, Retrieved 24 July, 2019
- Energy-fundamentals, energy: uni-leipzig.de, Retrieved 25 August, 2019
- Assessment-of-power-coefficient-of-an-offline-wind-turbine-generator-system-IJERTV2IS90929: ijert.org, Retrieved 26 January, 2019

# Environmental Impacts of Wind Energy

Wind power can have adverse impacts on humans, wildlife and the environment. It includes degradation of the habitat of wildlife, plants and fish, threat to flying animals, noise pollution, visual impacts on landscape, etc. This chapter has been carefully written to provide an easy understanding of these impacts of wind power.

Operation of wind power has zero emissions of harmful substances. It does not add to global warming, the "fuel" is free, and is quite evenly distributed around the world. The energy needed to produce and install the turbine amounts to three months of turbine production. But, as with other sources of energy, wind power does have an environmental impact. The impact on wildlife is likely low compared to other forms of human and industrial activity. However, negative impacts on certain populations of sensitive species are possible, and efforts to mitigate these effects should be considered in the planning phase. wind energy, like any other industrial activity, may cause impacts on the environment which should be analyzed and mitigated.

## Environmental Benefits

What are the advantages to the environment that is caused by wind energy?

Primarily, wind energy do not cause water or air emissions, and do not produce any kind of hazardous waste as well. Moreover, wind power does not make use of natural resources like oil, gas or cause and therefore will not cause damage to the environment through resource transportation and extraction and also do not need consequent amounts of water during operation.

Wind energy is not only a favorable electricity generation technology that reduces emissions (of other pollutants as well as $CO_2$, $SO_2$ and $NO_x$), it also avoids significant amounts of external costs of conventional fossil fuel-based electricity generation.

More and more use of wind energy should be made in order to prevent the problem of global warming.

Wind energy plants are considered a green power technology because it has only minor impacts on the environment. Wind energy plants produce no air pollutants or greenhouse gases.

Wind energy is an ideal renewable energy because:

- It is a pollution-free, infinitely sustainable form of energy.
- It doesn't require fuel.
- It doesn't create greenhouse gases.
- It doesn't produce toxic or radioactive waste.

## Disadvantages

Any means of energy production impacts the environment in some way, and wind energy is no different. Like every other energy technology, wind power plants do have some effects on the environment. Wind turbines cause virtually no emissions during their operation and very little during their manufacture, installation, maintenance and removal. Compared to the environmental impact of traditional energy sources, the environmental impact of wind power is relatively minor.

Wind farms are often built on land that has already been impacted by land clearing. The vegetation clearing and ground disturbance required for wind farms is minimal compared with coal mines and coal-fired power stations. If wind farms are decommissioned, the landscape can be returned to its previous condition.

The major challenge to using wind as a source of power is that the wind is intermittent and it does not always blow when electricity is needed. Wind energy cannot be stored (unless batteries are used); and not all winds can be harnessed to meet the timing of electricity demands.

Good wind sites are often located in remote locations, far from cities where the electricity is needed. Wind resource development may compete with other uses for the land and those alternative uses may be more highly valued than electricity generation. Although wind power plants have relatively little impact on the environment compared to other conventional power plants, there is some concern over the noise produced by the rotor blades, aesthetic (visual) impacts, and sometimes birds have been killed by flying into the rotors. Most of these problems have been resolved or greatly reduced through technological development or by properly sitting wind plants.

To the extent that we understand how, when, and where wind-energy development most adversely affects organisms and their habitat, it will be possible to mitigate future impacts through careful sitting decisions.

## Ecological Impacts

There are two major ways that wind-energy development may influence ecosystem structure and functioning— through direct impacts on individual organisms and through impacts on habitat structure and functioning. Environmental influences of wind-energy facilities can propagate across a wide range of spatial scales, from the location of a single turbine to landscapes, regions, and the planet, and a range of temporal scales from short-term noise to long-term influences on habitat structure and influences on presence of species. The ecological influences of wind-energy facilities are complex, and can vary with spatial and temporal scale, location, season, weather, ecosystem type, species, and other factors. Moreover, many of the influences are likely cumulative and ecological influences can interact in complex ways at wind energy facilities and at other sites associated with changed land-use practices and other anthropogenic disturbances. Wind turbines cause fatalities of birds and bats through collision, most likely with the turbine blades. Species differ in their vulnerability to collision, in the likelihood that fatalities will have large scale cumulative impacts on biotic communities, and in the extent to which their fatalities are discovered. the data are inadequate to assess relative risk to passerines and other small birds. It is possible that as turbines become larger and reach higher, the risk to the more abundant bats and nocturnally migrating passerines at these altitudes will increase. Determining the effect of turbine size on avian risk will require more data from direct comparison of fatalities from a range of turbine types. The construction and maintenance of wind-energy facilities also alter ecosystem structure through vegetation clearing, soil disruption and potential for erosion, and noise. Alteration of vegetation, including forest clearing, represents perhaps the most significant potential change through fragmentation and loss of habitat for some species. Changes in forest structure and the creation of openings alter microclimate and increase the amount of forest edge. Plants and animals throughout an ecosystem respond differently to these changes. There might also be important interactions between habitat alteration and the risk of fatalities, such as bat foraging behavior near turbines.

Standardized studies should be conducted before sitting and construction and after construction of wind-energy facilities to evaluate the potential and realized ecological impacts of wind development. Pre-sitting studies should evaluate the potential for impacts to occur and the possible cumulative impacts in the context of other sites being developed or proposed. Likely impacts could be evaluated relative to other potentially developable sites or from an absolute perspective. In addition, the studies should evaluate a selected site to determine whether alternative facility designs would reduce potential environmental impacts. Post-construction studies should focus on evaluating impacts, actual versus predicted risk, causal mechanisms of impact, and potential mitigation measures to reduce risk and reclamation of disturbed sites.

## Impacts on Humans

The human impacts include aesthetic impacts; impacts on cultural resources, such as historic, sacred, archeological, and recreation sites; impacts on human health and wellbeing, specifically from noise and from shadow flicker; economic and fiscal impacts; and the potential for electromagnetic interference with television and radio broadcasting, cellular phones, and radar. This is not an exhaustive list of all possible human impacts from wind-energy projects.

## Cultural Impacts

Wind-energy facilities create both positive and negative recreational impacts. On the positive side, many windenergy projects are listed as tourist sights: some offer tours or provide information areas about the facility and wind energy in general; and several are considering incorporating visitor centers. There are two types of potential negative impacts on recreational opportunities: direct and indirect. Direct impacts can result when existing recreational activities are either precluded or require rerouting around a wind-energy facility. Indirect impacts include aesthetic impacts (addressed above) that may affect the recreational experience. These impacts can occur when scenic or natural values are critical to the recreational experience. In analyzing impacts on historic, sacred, and archeological sites, the primary concern is that no permanent harm should be done that would affect the integrity of the site. Whether or not a wind-energy project would damage the resource may depend on the specific nature of the historic resources involved.

Unlike housing developments, wind-energy projects cannot be screened from view, except behind intervening topography and vegetation. Such issues are likely to arise as wind projects are proposed in cultural landscapes, and guidance as to what constitutes an undue impact to historic or sacred sites and areas will be necessary.

## Impacts on Human Health

Wind-energy projects can have positive as well as negative impacts on human health and wellbeing. The positive impacts accrue mainly through improvements in air quality, as discussed previously in this report. These positive impacts (i.e., benefits) to health and well-being are diffuse; they are experienced by people living in areas where conventional methods of electricity generation are used less because wind energy can be substituted in the regional market. In contrast, to the extent that wind-energy projects create negative impacts on human health and wellbeing, the impacts are experienced mainly by people living near wind turbines that are affected by noise and shadow flicker.

## Local Economic Impacts

Wind-energy projects can have a range of economic and fiscal impacts, both positive and negative. Some of those impacts are experienced at the national or regional level; these involve, and for example, tax credits and other monetary incentives to encourage wind-energy production, as well as effects of wind energy on regional energy pricing. In this topic, the focus is on the local level: on private economic impacts, positive and negative, as well as on public revenues and costs.

## Visual Impact

Landscape perceptions and visual impacts are key environmental issues in determining wind farm applications related to wind energy development as landscape and visual impacts are by nature subjective and changing over time and location.

The characteristics of wind developments may cause landscape and visual effects. These characteristics include the turbines (size, height, number, material and color), access and site tracks, substation buildings, compounds, grid connection, anemometer masts, and transmission lines. Another characteristic of wind farms is that they are not permanent, so the area where the wind farm has been located can return to its original condition after the decommissioning phase. While visual impact is very specific to the site at a particular wind farm, several characteristics in the design and sitting of wind farms have been identified to minimize their potential visual impact.

## Noise Impact

Noise from wind developments has been one of the most studied environmental impacts of this technology. Noise, compared to landscape and visual impacts, can be measured and predicted fairly easily.

As with any machine involving moving parts, wind turbines generate noise during operation. Noise from wind turbines arises mainly from two sources: 1) mechanical noise caused by the gearbox and generator; and 2) aerodynamic noise caused by interaction of the turbine blades with the wind

Experience acquired in developing wind farms suggests that noise from wind turbines is generally very low. The comparison between the number of noise complaints about wind farms and about other types of noise indicates that wind farm noise is a small-scale problem in absolute terms. Information from the US also suggests that complaints about noise from wind projects are rare and can usually be satisfactorily resolved.

## Impact on Land Use

National authorities consider the development of wind farms in their planning policies for wind energy projects.

Decisions on sitting should be made with consideration to other land users. Regional and local land-use planners must decide whether a project is compatible with existing and planned adjacent uses, whether it will modify negatively the overall character of the surrounding area, whether it will disrupt established communities, and whether it will be integrated into the existing landscape. Land use planning rules in some countries recommend avoiding areas with protected designations; in others, specific areas have been earmarked for potential wind farm development.

## Reduce the Negative Environmental Impacts of Wind Energy

The negative environmental impacts from wind energy installations are much lower in intensity than those produced by conventional energies, but they still have to assessed and mitigated when necessary. There are specific conditions that must be in place before an area can be considered suitable for a wind farm development. These conditions include factors such as: wind climate, topographical, logistical and ecological constraints.

A strategic environmental assessment (SEA) is the procedure used to evaluate the adverse impacts of any plans and programs on the environment. National, regional and local governments must undertake SEAs of all wind energy plans and programs that have the potential for significant environmental effects. The ecological influences of wind-energy facilities are complex, and can vary with spatial and temporal scale, location, season, weather, ecosystem type, species, and other factors. Moreover, many of the influences are likely cumulative and ecological influences can interact in complex ways at wind energy facilities and at other sites associated with changed land-use practices and other anthropogenic disturbances. Because of this complexity, evaluating ecological influences of wind-energy development is challenging and relies on understanding factors that are inadequately studied. Despite this, several patterns are beginning to emerge from the information currently available. Increased research using rigorous scientific methods will be critical to filling existing information gaps and improving reliability of predictions.

In conclusion, we must decide that if we have to produce electricity, it is certainly preferable to produce it in a way which has the smallest possible impact on the environment. From a technical and economic standpoint, the most mature form of renewable and "clean" energy is wind energy. It can effectively contribute to combating climate change while at the same time providing various environmental, social and economic benefits. On the other hand, it is necessary to minimize the impact of the wind energy, particularly in terms of environment (preservation of protected areas) and human health (noise and visual impact).

Harnessing power from the wind is one of the cleanest and most sustainable ways to generate electricity as it produces no toxic pollution or global warming emissions. Wind is also abundant, inexhaustible, and affordable, which makes it a viable and large-scale alternative to fossil fuels.

Despite its vast potential, there are a variety of environmental impacts associated with wind power generation that should be recognized and mitigated.

## Land Use

The land use impact of wind power facilities varies substantially depending on the site: wind turbines placed in flat areas typically use more land than those located in hilly areas. However, wind turbines do not occupy all of this land; they must be spaced approximately 5 to 10 rotor diameters apart (a rotor diameter is the diameter of the wind turbine blades). Thus, the turbines themselves and the surrounding infrastructure (including roads and transmission lines) occupy a small portion of the total area of a wind facility.

A survey by the National Renewable Energy Laboratory of large wind facilities in the United States found that they use between 30 and 141 acres per megawatt of power output capacity (a typical new utility-scale wind turbine is about 2 megawatts). However, less than 1 acre per megawatt is disturbed permanently and less than 3.5 acres per megawatt are disturbed temporarily during construction. The remainder of the land can be used for a variety of other productive purposes, including livestock grazing, agriculture, highways, and hiking trails. Alternatively, wind facilities can be sited on brownfields (abandoned or underused industrial land) or other commercial and industrial locations, which significantly reduces concerns about land use.

Offshore wind facilities require larger amounts of space because the turbines and blades are bigger than their land-based counterparts. Depending on their location, such offshore installations may compete with a variety of other ocean activities, such as fishing, recreational activities, sand and gravel extraction, oil and gas extraction, navigation, and aquaculture. Employing best practices in planning and siting can help minimize potential land use impacts of offshore and land-based wind projects.

## Wildlife and Habitat

The impact of wind turbines on wildlife, most notably on birds and bats, has been widely document and studied. A recent National Wind Coordinating Committee (NWCC) review of peer-reviewed research found evidence of bird and bat deaths from collisions with wind turbines and due to changes in air pressure caused by the spinning turbines, as well as from habitat disruption. The NWCC concluded that these impacts are relatively low and do not pose a threat to species populations.

Additionally, research into wildlife behavior and advances in wind turbine technology have helped to reduce bird and bat deaths. For example, wildlife biologists have found that bats are most active when wind speeds are low. Using this information, the Bats and Wind Energy Cooperative concluded that keeping wind turbines motionless during times of low wind speeds could reduce bat deaths by more than half without significantly affecting power production. Other wildlife impacts can be mitigated through better siting of wind turbines. The U.S. Fish and Wildlife Services has played a leadership role in this effort by convening an advisory group including representatives from industry, state and tribal governments, and nonprofit organizations that made comprehensive recommendations on appropriate wind farm siting and best management practices.

Offshore wind turbines can have similar impacts on marine birds, but as with onshore wind turbines, the bird deaths associated with offshore wind are minimal. Wind farms located offshore will also impact fish and other marine wildlife. Some studies suggest that turbines may actually increase fish populations by acting as artificial reefs. The impact will vary from site to site, and therefore proper research and monitoring systems are needed for each offshore wind facility.

## Public Health and Community

Sound and visual impact are the two main public health and community concerns associated with operating wind turbines. Most of the sound generated by wind turbines is aerodynamic, caused by the movement of turbine blades through the air. There is also mechanical sound generated by the turbine itself. Overall sound levels depend on turbine design and wind speed.

Some people living close to wind facilities have complained about sound and vibration issues, but industry and government-sponsored studies in Canada and Australia have found that these issues do not adversely impact public health. However, it is important for wind turbine developers to take these community concerns seriously by following "good neighbor" best practices for siting turbines and initiating open dialogue with affected community members. Additionally, technological advances, such as minimizing blade surface imperfections and using sound-absorbent materials can reduce wind turbine noise.

Under certain lighting conditions, wind turbines can create an effect known as shadow flicker. This annoyance can be minimized with careful siting, planting trees or installing window awnings, or curtailing wind turbine operations when certain lighting conditions exist.

The Federal Aviation Administration (FAA) requires that large wind turbines, like all structures over 200 feet high, have white or red lights for aviation safety. However, the FAA recently determined that as long as there are no gaps in lighting greater than a half-mile, it is not necessary to light each tower in a multi-turbine wind project. Daytime lighting is unnecessary as long as the turbines are painted white.

When it comes to aesthetics, wind turbines can elicit strong reactions. To some people, they are graceful sculptures; to others, they are eyesores that compromise the natural landscape. Whether a community is willing to accept an altered skyline in return for cleaner power should be decided in an open public dialogue.

## Water Use

There is no water impact associated with the operation of wind turbines. As in all manufacturing processes, some water is used to manufacture steel and cement for wind turbines.

## Life-cycle Global Warming Emissions

While there are no global warming emissions associated with operating wind turbines, there are emissions associated with other stages of a wind turbine's life-cycle, including materials production, materials transportation, on-site construction and assembly, operation and maintenance, and decommissioning and dismantlement.

Estimates of total global warming emissions depend on a number of factors, including wind speed, percent of time the wind is blowing, and the material composition of the wind turbine. Most estimates of wind turbine life-cycle global warming emissions are between 0.02 and 0.04 pounds of carbon dioxide equivalent per kilowatt-hour. To put this into context, estimates of life-cycle global warming emissions for natural gas generated electricity are between 0.6 and 2 pounds of carbon dioxide equivalent per kilowatt-hour and estimates for coal-generated electricity are 1.4 and 3.6 pounds of carbon dioxide equivalent per kilowatt-hour.

Wind Power and the Environment.

Wind power is a clean and environmentally friendly technology that catches or extracts the kinetic energy in the wind and converts it to other useful forms of energy, for example, mechanical work to pump water or grind corn, or electrical energy to power our homes.

Compared to other more traditional forms of energy sources, the environmental impact of wind power on our planet, either local, regional or global, can be viewed as being minor, since being a renewable energy source, wind power does not pollute the atmosphere or consume any fossil fuels when operating because no fuel is burned. Nor does it consume clean water for cooling or leave any hazardous waste behind that could potentially harm the environment.

Although wind turbines produce clean energy and omit virtually zero emissions during their operation, like many other industrial activities, wind energy may cause visual and sound impacts on the environment as well as a danger of collisions to birds and bats from the blades of these towering structures.

The wind is a clean, free and inexhaustible resource being nothing more than the air in motion. Since this air has a mass, as it moves it stores large amounts of kinetic energy.

Then the wind has a lot of power and a wind turbines blades can turn this power into mechanical work or electricity power. The energy content of the wind increases with height but how large this increase will obviously depend on the type of the terrain. Flat lands or mountainous areas.

Clearly then, the higher the turbines tower the faster the wind speed over the blades which also allows for longer blades and faster tip speeds, increasing efficiency. However, the footprint required for a single wind turbine and the additional spacing required between wind turbines can be large increasing the land usage and visual impact of the wind farm.

Then we can see that whilst wind energy is a clean and green technology offering many positive and environmental benefits, it is not completely free of impacts on the environment with some of the main issues being.

## Environmental Impacts of the Life Cycle

Construction – The component parts of wind turbines, the tower, nacelle, hub, blades, etc, need to be manufactured and the raw materials used during their construction such as steel, aluminium, copper cable, glass fibre, etc all needs to be sourced, processed and transported to the manufacturing facility adding to the environmental footprint of the turbine.

Transportation and On-site Installation – As well as transportation of the raw materials, once built, wind turbines need to be transported using large lorries and trucks along the road network, possibly over some distance, to the site for installation. Construction of access roads and bridges for these heavy loads through fields and local countryside to the point of installation is generally required.

Large concrete foundations and trenches for grid connection cables can create its own environmental issues. The work of on-site erection and assembling requires large cranes and a team of people with vans burning fossil fuels adding to the environmental footprint.

Operation – Once erected and working, periodic maintenance of the turbines, including inspection, lubrication and gearbox oil change, repairs, and cleaning is required plus the transportation of these maintenance materials and crews, usually by van or truck. Also the recycling and disposing of used parts and lubricants could affect the environment if not done correctly.

Dismantling After Service – Nothing lasts forever, so once the wind turbine has reached its end-of-service life, the work of dismantling the wind turbine and the transportation, again by lorries and trucks from the wind farm site to the final disposal site adds to the overall environmental impact of the wind turbine.

## Visual Impact on the Environment

The direct physical impact of wind power on the environment mainly consists of the concrete foundations, access road and electrical cables to the grid, as well as the turbine itself, which will require a certain amount of air space around its blades, the biggest concern for most people with regards to wind power and the environment is the visual impact.

Wind turbines are man-made vertical steel structures with large rotating blades, and therefore visually have an impact on the local landscape, just like most other structures, large buildings, factories, power grid pylons, main roads, etc. Since wind turbines are tall and have rotating blades, they have the potential of attracting people's attention. Thus turbines are perceived to have a relatively large impact on both the landscape and distant horizon.

The number and size of wind turbines and their positioning plays an important role in how the public view turbines and wind farms. If the wind turbines are positioned and spread out along a line, the land requirement could be very small but they appear to spread along the horizon. If they are arranged together in groups or in a matrix the land requirement may increase becoming dominant points on the landscape.

Then in order to maintain public acceptance, the siting of wind turbines and wind farms need to be designed in such a way as to minimise the various visual and amenity impacts when viewed from both short and long distances allowing over time for wind farms to be better accepted. To some people, wind farms enhance the beauty of the area.

Also, the perception that these large areas of turbines use up valuable farmland can sometimes be farther from the truth. While the turbines themselves may cover a large area of farmland, most of the land they stand on can be used as before, as arable or even pasture land as only the small areas on which the turbine foundations stand is unavailable for use.

## Sound Emissions

Offshore wind farms located in the seas and estuary's are too far away from houses and buildings to be affected by the noise generated by the rotating turbines. Nowadays, most modern vertical wind turbine designs have improved significantly to the point where any mechanically generated noises from the operation of the gearbox, generator and other moving parts has been virtually eliminated through the use of good design and insulating materials in the nacelle. Then the issue of noise emissions is one of aerodynamic noise and vibrations from the rotating blades.

The aerodynamic noise produced by the rotation of the turbine blades generates a swishing sound as a result of the tip speed. The design of modern turbine blades reduces this aerodynamic noise by decreasing the rotational speeds of the blade tip or by using better blade pitch control in high winds.

The aerodynamic sound emissions created by a wind turbine increases as the wind speed increases. However as the wind speed increases, the noise from the rotating turbines is usually masked by other background noise such as the movement of trees or if sited near an industrial or urban area. Turbine sound emissions will naturally decrease as the distance from the listener to the wind turbines increases.

## The Impact on Bird Life

Wind turbines are tall vertical structures with long rotating blades that may represent a risk of collision to the birds which get too close. Most ornithologist agree that it is difficult to reach a definite conclusion about the impact that wind turbines have on bird mortality but most seem to agree that larger birds of prey are affected more.

Bird collisions with turbine blades has been an issue for many years at some older wind farm sites due to poor siting, multi-bladed designs and open tower technology. Studies around the world has shown that bird fatalities may be as low as a few per turbine per year compared to the thousands that die each year from colliding with buildings, power lines, vehicles, etc.

When put into context, the human impact on bird life from loss of habitat, industrialisation, over exploitation of their natural environment, hunting, pollution, etc. represents a far greater threat to the worlds bird population than collisions with turbine blades. Each year more and more bird species face extinction.

## The Impact on Flora and Fauna

The environmental impact of wind power on flora and fauna depends on the types of vegetation and animal life within the area of the turbines or wind farm. While there has been many studies around the world on the risk of impacts on birds, little has been done about the wildlife living beneath and around the wind farms and turbines.

The local landscape, vegetation and habitat of many small animals and insects can be affected during the construction phase as access roads, cable ditches and foundations change the area.

However, the construction process may take only a few weeks, depending on the site conditions and project size, but once complete, landscaping and agricultural activity can continue right up to the turbine bases. Then wind turbines may not make such an impact on the environment after all.

## Wind Power and the Environment

Wind turbines are tall visible objects and as such have a visual impact on their surrounding landscape, like most other structures. But all is not bad bad either for wind power and the environment. Surveys and opinion polls have shown that public attitudes towards the installation of wind turbines around the countryside has shown consistent, strong support for this sustainable and renewable energy technology. Wind power is more popular now than ever before.

The impact of wind power on the environment can be greatly reduced with proper planning, design and implementation. The environmental benefits of wind energy is clear as wind turbines create virtually no emissions during their continuous operation and also very little during their manufacture, installation, maintenance and final removal.

The economic lifetime of a wind turbine may be at least 20 years, which can be prolonged by annual maintenance, but once de-commissioned and dismantled, the site could be restored back to its original state with most of the turbines components being recyclable.

Wind energy is a clean and renewable energy source because the fuel (the wind) is free. So wind generated electrical energy should be used as much as possible to replace that generated from coal and gas. Then wind energy has a major role to play in opposing climate change by reducing the amount of $CO_2$ emissions produced from conventional fossil fuel power generation.

Around the world, hundreds of thousands of turbines have already been installed on available land, but as it becomes more difficult to find good open sites to erect more wind turbines planners are looking out to sea. Large offshore wind farms have already been developed as the wind resource offshore in the sea is far better than on land increasing power generation. However the installation cost are also much higher.

While the visual and sound impacts or the availability of land may not be a problem offshore, the environmental affects on fish and other marine mammals can be a positive one. While the construction phase may disturb many of the fish species, the creation of new artificial habitats around the turbine bases have had significant effects on fish population. With a ban on shipping (other than maintenance ships) and fishing around the offshore turbines, local fish species have been able to develop. This availability of food may even attract new species of fish and subsequently marine mammals.

As we have seen here that the impact of wind power and the environment can be positive or negative depending on your outlook. The energy output of a wind turbine over its operational life is far greater than the energy used to construct, operate and eventually dismantle the turbine. The visual impact of wind turbines on the environment is very specific to the particular site but on average the vast majority of the public are now in favour of wind energy showing a consistently high level of support.

Clearly from an environmental point of view, the generation of energy using wind power is the best option. The environmental risks associated with climate change, greenhouse gases and the impact of acid rain on agriculture, forests, lakes and human health as a direct result of the combustion of fossil fuels is greatly decrease as more electricity is being generated by the wind.

# PERMISSIONS

All chapters in this book are published with permission under the Creative Commons Attribution Share Alike License or equivalent. Every chapter published in this book has been scrutinized by our experts. Their significance has been extensively debated. The topics covered herein carry significant information for a comprehensive understanding. They may even be implemented as practical applications or may be referred to as a beginning point for further studies.

We would like to thank the editorial team for lending their expertise to make the book truly unique. They have played a crucial role in the development of this book. Without their invaluable contributions this book wouldn't have been possible. They have made vital efforts to compile up to date information on the varied aspects of this subject to make this book a valuable addition to the collection of many professionals and students.

This book was conceptualized with the vision of imparting up-to-date and integrated information in this field. To ensure the same, a matchless editorial board was set up. Every individual on the board went through rigorous rounds of assessment to prove their worth. After which they invested a large part of their time researching and compiling the most relevant data for our readers.

The editorial board has been involved in producing this book since its inception. They have spent rigorous hours researching and exploring the diverse topics which have resulted in the successful publishing of this book. They have passed on their knowledge of decades through this book. To expedite this challenging task, the publisher supported the team at every step. A small team of assistant editors was also appointed to further simplify the editing procedure and attain best results for the readers.

Apart from the editorial board, the designing team has also invested a significant amount of their time in understanding the subject and creating the most relevant covers. They scrutinized every image to scout for the most suitable representation of the subject and create an appropriate cover for the book.

The publishing team has been an ardent support to the editorial, designing and production team. Their endless efforts to recruit the best for this project, has resulted in the accomplishment of this book. They are a veteran in the field of academics and their pool of knowledge is as vast as their experience in printing. Their expertise and guidance has proved useful at every step. Their uncompromising quality standards have made this book an exceptional effort. Their encouragement from time to time has been an inspiration for everyone.

The publisher and the editorial board hope that this book will prove to be a valuable piece of knowledge for students, practitioners and scholars across the globe.

# INDEX

www.ingramcontent.com/pod-product-compliance
Lightning Source LLC
Chambersburg PA
CBHW061247190326
41458CB00011B/3604